S0-BMA-274

A Chanticleer Press Edition

Photographs Assembled and
Edited by Milton Rugoff and Ann Guilfoyle
Design and Coordination by
Massimo Vignelli and Gudrun Buettner

Wild Creatures

A Pageant of the Untamed

Photographs by:
Anthony Bannister,
Jen & Des Bartlett, Erwin Bauer,
René Pierre Bille, Les Blacklock,
Dennis Brokaw, Bill Browning,
Fred Bruemmer, Robert Carr,
Patricia Caulfield,
David Cavagnaro,
Glenn D. Chambers,
Neville Coleman, Ed Cooper,
William R. Curtsinger,
Thase Daniel, Edward R. Degginger,
Jack Dermid, Tui De Roy,
John Dominis, Hans Dossenbach,
Douglas Faulkner,
Jean-Louis Frund,
Jessie O'Connell Gibbs, Fritz Goro,
Clem Haagner, Bob Harrington,
George Holton, Maurice Hornocker,
Eric Hosking, Philip Hyde,
M. Philip Kahl, Hubertus Kanus,
S. D. MacDonald,
Lorus and Margery Milne,
Oxford Scientific Films,
Willis Peterson, David Plowden,
Hans Reinhard, Co Rentmeester,
Edward S. Ross, Galen Rowell,
Leonard Lee Rue III,
Emil Schulthess, George Silk,
Gordon S. Smith,
Charles Steinhacker, Harald Sund,
Karl H. Switak, Valerie Taylor,
Karl Weidmann, Larry West,
Steven C. Wilson

Text by Franklin Russell

Simon and Schuster, New York

Preceding page
A giant tortoise faces a chill dawn on the rim of an active volcano, Alcedo, in the Galapagos Islands. This sulphur-scented island harbors giant tortoises, which may live a century or more, grow to 600 pounds, and reach six feet in length. Because of their island isolation, several species of these remarkable creatures evolved from a race of giant tortoises that once wandered open areas of the mainland. Now the tortoises of the Galapagos and their relatives living on Aldabra atoll in the Indian Ocean are in danger of extinction.
(Tui De Roy)

Published by Simon and Schuster
Rockefeller Center, 630 Fifth Avenue
New York, New York 10020

SBN 671-22098-5
Library of Congress Catalog Card
Number 75-4417
Planned and produced by Chanticleer
Press, Inc., New York
Manufactured in Milan, Italy, by
Amilcare Pizzi, S.p.A.
1 2 3 4 5 6 7 8 9 10

Chanticleer Staff
Publisher: Paul Steiner
Editor: Milton Rugoff
Picture Editor: Ann Guilfoyle
Production: Gudrun Buettner,
Helga Lose, and Judy Shugerman
Design: Massimo Vignelli

Contents

Introduction

Preceding page
The turbulence of the sea made it possible for the early forms of marine life to migrate, voluntarily or involuntarily, to the furthest limits of the earth. The plants and animals traveled both horizontally and vertically, carried by waves, upwellings, and submarine currents, and colonized most of the available ecological niches in the world's oceans. (Harald Sund)

The first creatures of the earth appeared in the sea; they crawled up the shore; they moved inland to occupy every part of the world. The simple history of life can be compressed into a sentence. But it is not so easy to compress the complicated process by which creatures became dependent on each other, on the plants around them, and on the places in which they lived.

Wherever each animal found its home—in stygian cave pools or buried deep inside elephants' bodies—its life was governed by its place. The leopard of the rain forest is not the same animal that hunts the savanna of eastern Africa. The salmon which swims up the Norwegian fjord is different from the fish sculling up the Skeena River in British Columbia, and both are unlike the salmon trapped forever in lakes.

For all its creatures, the earth is a series of distinct worlds, each self-contained and self-perpetuating, though joined like a chain to one another.

I have watched animals in many places; a nocturnal parrot searching for grubs in the moonlight; a tiny, tree-climbing wallaby leaping nimbly into the branches of a coolibah tree; a skua flying rapidly near the South Pole; a willow ptarmigan in Labrador pecking at my foot in an attempt to protect her chicks. They live in widely separated environments, yet all are interdependent, in one way or another.

These creatures are also part of me. The chemistry of my blood is similar to theirs. And our blood is linked, in turn, to the chemical mix present in ancient sea creatures. The cells in our bodies resemble those of trees and grass. In watching these living things, I see a remote part of myself.

When I travel, other creatures are traveling too, and this gives me an illusion of the similarity between tamed and untamed lives. The arctic tern speeds along the coast of Greenland, or Africa, or Antarctica, and it is the only creature, other than the killer whale, to have mastered both polar worlds. But then notions of similarity die when I find hundreds of exhausted swallows flopping on rocks high in the Swiss Alps. An early cold spell has brought them down before they could reach Italy. Their travel is not merely dangerous; it may kill them all. At midnight, I hear the high-pitched twittering of songbirds flying over the Bay of Fundy toward the Gulf of St. Lawrence, and know they are heading for Cape Breton, Newfoundland, Quebec, and Labrador. I see streaming flights of monarch butterflies in New Hampshire and Wisconsin.

How do they find their way? I can explain the warblers, and perhaps the terns and swallows, as being capable of reading celestial signs, and sensing the magnetic fields of earth. But what of the butterflies? Some of them reach the Gulf of Mexico. What signals do they receive to guide them?

I am linked to these creatures by my chemistry, by the nature of my cells, the functioning of my lungs and eyes, my sense of taste and smell. That is simple enough. What is not so simple is that to migrate over such great distances, the creatures must be connected to some kind of electrical signaling system. The only electrical signals received on earth come from outer space, from distant stars. If this is the system they use, then interdependence becomes a cosmic word.

But the only certitudes are the earth's environments, and it is no accident that they influence both humanity and wild creatures in similar ways. The song thrush, pouring out its autumn song from an English hedgerow, is surely singing for pleasure, or at least I choose to believe so. The otter, whisking down his mud slide, surely knows the meaning of enjoyment. We have come to believe that anthropomorphizing animals—attributing to them human emotions and motivations— is wrong. But to me, this is nonsense. Animals have feelings, and separate and distinct personalities which are as various and surprising as anything in the human sphere.

And after all, the purpose of animal life is the same as human life. It is survival. All of us seek stability within our worlds and it does not matter how imperfectly realized is the ideal. The lemmings die by the millions in their suicidal migrations. Virus diseases strike down wildebeest in Africa. Drought decimates the kangaroos in Australia. But these kinds of disasters are unimportant. The vital fact is the preservation of the environment for the next generation of animals.

From this comes another unifying truth for most animals. When I see the cheetah stretch its body in that last blinding burst of speed that should send its frantic victim spinning, only to have the cat trip and fall in a spectacular stumble, I understand how this truth applies to all life. Nothing is available for free. Every advantage won from evolution has been paid for by drawbacks, some of them crippling, in the present. The elephant destroys the forest but a tiny forest creature may cripple the elephant. The tsetse fly gorges on blood and is then destroyed by others who feed on the blood it has stolen from others. There is no perfect state where struggle and tension can be banished. The tension, in fact, may resemble the taut stringing of a tennis racket. If the stringing goes limp, the ball cannot be hit accurately, or hard. The importance of the condition of the environment is partially understood today, but its capacity to produce distinct characteristics in living things is critical, and most likely shapes the quality of all life. Birds isolated on islands without conventional predators may greet men by alighting on the barrels of their guns, or settling on their heads or hats. The creature in its world becomes only as wary, as tough, as agile, as smart, as it needs to be to survive. The great auk did not need to fly, and it stopped doing so, and thus became extinct when men in boats arrived at its island breeding grounds. Other sea birds, relying on wings to fight the gales and blizzards, the heat and cold of oceanic wastes, remained tough enough to withstand the pressure of the hunting men.

Emperor penguins choose winter to breed on blizzard-swept ice in Antarctica. Geese fly over Mount Everest. Tiny insects thrive in high mountain snow. Lichens flourish on as little as a couple of day's growth a year in some icy places. The dark depths of the oceans have been colonized by light-producing creatures. The driest deserts hold life.

In watching the wild creatures, we might understand that the state of the environment is not merely vital to survival; it is the supreme arbiter. When it is in balance and stable, then so is all life.

1 The Prolific Seas

Photographs by:
Fritz Goro: Time/Life Picture Agency, 13
Oxford Scientific Films, 14
Ron and Valerie Taylor, 16, 18, 24
Douglas Faulkner, 19, 20, 26
Neville Coleman: Tom Stack & Associates, 22
William R. Curtsinger, 28

The sunsets of the Indian Ocean are fabled; the sun appears unbelievably enlarged, and the sea turns vermilion. Once, during such a sunset, I watched scores of flying fish break the red surface and speed toward the sun. It looked as though they were hurling themselves into fire. Then, one by one, they splashed back into the crimson water. The sun sank. Dusk stole quickly over the ship on which I was traveling. For me, the flying fish evoked a truly cosmic image of life: fragile creatures set against a universal power and surviving in its grip.

The sea has developed the diatom, a pinhead-sized globe of silica which contains a single, pulsing cell; and the sea has produced the one-hundred-ton whale, which eats millions of diatoms in a single meal. It is a world where almost anything is possible. Life began in the sea. It still contains about 90 per cent of all living things. It is the world's thermostat, adjusting temperatures everywhere; it dispenses typhoons and doldrums, regulates currents, decides the fate of civilizations.

The swelling of life in nearly all the seas and oceans of the world takes millions of different forms. The teeming plankton, particularly the immense proliferations of diatoms and other tiny plants, give sustenance to their animal hunters. Countless billions of protoplasmic blobs of life swarm in the vanguard of huge populations of shrimp and shrimplike creatures.

The sea produces life prolifically in large measure because it is so plastic. Movement is possible in almost any direction. Codfish rise from winter depths to hunt in summer shallows. Tuna migrate thousands of miles. Plankton rise and fall in migrations of hundreds of feet every day. The plastic sea allows its creatures freedom to move toward the sources of food, or to escape approaching enemies. It insures a universal mix, making the sea a kind of planetary chemical experiment in which the fundamental ingredients necessary for life—oxygen, phosphorus, potassium, nitrogen, calcium, carbon dioxide—are all contained in the vast beaker of oceanic water where they are influenced by the sun.

The sea absorbs both cold and heat, and the manner of their release dictates weather. If the chilly East Greenland Current, or the equally chilly Labrador Current, diminishes its flow, Europe benefits from an earlier spring and more bountiful crops. If the cold Humboldt Current, sweeping north up the western coast of South America, warps its flow a few miles offshore, Chile's highlands are devastated by torrential rains. Billions of fish die, and entire populations of sea birds perish. Conversely, the warm Gulf Stream allows palms to grow in southern Ireland.

The fluidity of the sea makes its chemicals available everywhere. Within its 330 million cubic miles of water are an estimated 1,200 quadrillion tons of oxygen, 150 quadrillion tons of hydrogen, 15 billion tons of iron, and more than 6 million tons of gold. The figures are so large they become meaningless, but that is the truth of the sea.

In my own lifetime, I have traveled only its fringes, yet I am more conscious of its contrasts than I am of the earth's, despite the height of Everest or the heat of Death Valley. The polar seas, piled up before me in endless, shattered pans of glittering blue ice, are set against a memory of the Red Sea, water temperature 99 degrees, and the air so hot that birds would not fly.

The deepest part of the sea, the Mariana Trench near the Philippines, is 40,000 feet down. Creatures live there, in the trench, under twenty-two tons of weight for every square inch, and flash lights at one another to communicate.

If you travel the seas enough, you carry a kaleidoscope of images with you: squid leaping from a midnight sea and flying into your lights; squid piling ashore in inexplicable wrecks; grunion pouring out eggs onto California beaches, and herring-like capelin swarming to breed on northern beaches; a killer whale rising in a seal's breathing hole north of Greenland and seizing the seal in a lethal grip; manta rays weighing up to two tons hurling themselves twenty feet into the air and sailing briefly before explosive crashes.

The creatures of the sea have taken evolutionary paths into every part of the oceans, and have assumed every conceivable form to capitalize on the resources there. The clam under my feet on the Atlantic beach has buried itself in sand to dodge its enemies. The giant clam of the tropics, with its gorgeous decorated and sinuous shells, has become so big and heavy that no creature can kill and eat it. Instead, its shells and body are invaded by tiny hunters which have made their own adjustment to capitalize on a creature that cannot run away.

The nudibranchs, mollusks without shells, have expanded their gills to give them an advantage in drawing the maximum amount of oxygen out of the water. They have developed a bitter taste in their bodies, so that only the inexperienced would dare to eat them, and so a shell is not necessary.

Underwater, I extended my understanding of the sea. I watched crabs crawling into abandoned snail shells to conceal themselves from enemies. I saw transparent shrimp, vital organs pulsing in midwater; sea urchins colored royal purple, and auks swimming underwater, their wings pumping in pursuit of panicky fish. A small shark darts from a rock refuge, seizes a fish and disappears, leaving a cloud of scales.

All divisions of life on earth are present in the sea, from the simple protozoans and viruses, the multitude of worms, the great arthropod group of crabs and shrimps and insects, the starfish and sea urchins, the sea anemones, the corals and jellyfish, the sponges and plants, to the backboned fish and the aquatic mammals. From these groupings of animals comes an ineffable sense of the continuity of life. I look up from the sea bottom of an Australian reef and see the drifting tentacles of a jellyfish, rows of primitive red eyes fringing the bottom of the pulsating clear body. This jellyfish is almost identical to one which roamed the sea during the Cambrian period, 500 millions years ago. The sea, it seems, rewards its creatures with a relatively stable world, and invites them not to change too quickly. Many living fossils thrive there: horseshoe crabs, starfish, corals, sponges. The temperature of the sea is always more constant than that of the air of the land, and water supplies salts and minerals everywhere.

The sea is ever-moving, but this does not mean that its food is readily available to all its creatures. In fact, most must make use of extremely specialized forms or habits, or both, to secure their share of sustenance. The starfish has made a home for itself along many shores, and far offshore. Blind, except for one spot sensitive to light at the end of each arm, it can devastate shellfish beds when its numbers periodically increase greatly. It can produce new arms when it loses old ones. It can extrude its stomach into a shellfish and inject digestive juices into its victim so that the liquefied body can be absorbed through the tissues of the extruded stomach. It pulls open some victims, but it is more likely to use chemicals to paralyze them. It matters little that the starfish cannot walk, swim, or see.

To protect themselves against this kind of ingenious hunter, other sea creatures have developed equally technical defense equipment. The lobster and the crab try to balance armor and agility so that they can flee, or fight, or stand fast against attackers too weak to break up their shells. The apparently helpless blowfish can expand its body in a second into a spherical ball studded with needle spines, which either makes the fish too large to be swallowed, or makes swallowing it an act of suicide.

The sea has been called a vast vegetable soup, composed of countless simple plants, particularly diatoms. But diatoms only swarm in limited sections of the sea, where their work of photosynthesis and ingestion of minerals is the first step in creating food for the animals that eat them. The shapes of the diatoms, which are endlessly different, hint at the complexity of life. Many are like uncut gems, sparkling with the colors of the universe. Each diatom constructs a tiny house of silica, which it extracts from the water around it.

When layers of water at different temperatures within the sea overturn in the collision of currents, or when a storm brings up salts which have fallen to lower depths, the diatoms abound. Their incalculable numbers color thousands of square miles of the sea with their blue and green and yellow forms. From these teeming nations spring a living pyramid of sea creatures.

After the diatoms come countless other planktonic creatures, through ascending orders of complexity: creatures with lashing, whiplike tails; creatures that create light and limn the waves with eerie phosphorescence; the copepods in their uncounted billions, jerking their limbs like oars.

Such simple animals opened the evolutionary road to the land. In the sea were developed the mouth, the stomach, the nervous system, blood, eyes, brains, limbs—all before there was any life on land. In the sea, shells, plates, spines, and poisonous glands evolved for protection. The trilobite, which flourished during the Cambrian period and survived, pretty much unchanged, for 370 million years, illustrates the slow pace of the evolving life. But, as it hauled its shell over ancient sea bottoms, an armored scavenger, it led the way for other, more flexible lives to follow; shellfish sixteen feet long, with gas-filled flotation chambers to ease movement; the first clams, starfish, and corals. Eventually, the higher plants evolved to the point at which they could leave the sea and move inland. After them came small creatures from the shallows and estuaries. The gifts of the sea were distributed on the land.

The amorphous mass of the sea appears to be uniform, repetitive in many climates. But, in fact, its substance is extremely variable. The Sargasso Sea, circling in weed-choked eternity, is not much like the waters of the Grand Banks, off Newfoundland, where the water is clouded with marine life, or the pellucid waters of Samoa, where sunlight throws a bright ultramarine glow over the bottom a score of feet down.

The sea can be ugly, poisonous, dangerous, when red algae swarm and turn the water pink, and line the gills of shellfish, and make many sea creatures poisonous to eat.

It can also be beautiful, particularly in tropical waters where the striking colors of fish and shellfish are turned brilliant by the sun pouring through the clear waters. Among the white walls of tropical reefs, multihued fish flash away. Like the gorgeously striped harlequin fish of Australian waters, many of them are dressed in colors that seem to be for their own sake, although more likely they are warnings to hunters of foul-tasting flesh, or parts of camouflage techniques.

Colors make the sea bright in places, but it is form that is more exciting. The hurrying throngs of mackerel, cutting the water to foam in their efforts to escape prowling sharks, are miniature torpedoes. The shark, poised in its blue world like some intergalactic spaceship, is one of the most beautifully realized instruments of the kill. And such form is used by other, dissimilar creatures that have similar functions. The dolphin, a warm-blooded creature whose ancestors returned to the sea from land life millions of years ago, has taken a form similar to that of the shark. It resembles the high-speed tuna and many other fish whose ancestors never ventured from the sea.

The infinite plasticity of the sea encourages movement, whether within its waters or above them. The codfish, the tuna, the swordfish, the whales, and many others move thousands of miles every year, and some cross hemispheres. The Laysan albatross, wheeling along behind my ship, flies millions of miles during its lifetime. I once sailed for fourteen hours through a jellyfish swarm in the Pacific. A stiff prevailing wind was pushing them along and they would likely travel hundreds of miles together. Sea snakes travel for miles to gather in the warm seas of the south Pacific. Young eels, rising from the Sargasso, may take two years to reach rivers far to the east.

Bluefin tuna spawn in the Caribbean and, depleted, head north, replenishing their strength as they forge up through the Gulf Stream toward the eastern shores of Canada. Salmon congregate along the shores of Greenland, then radiate in migrations that take them to Europe and to the shores of the American coast. And some of the above-water migrations are the greatest of all, most notably the arctic tern in its flight south from subpolar regions to the southwest coast of Africa, and thence to Antarctica.

This freedom of movement makes the sea and its creatures dynamic and fascinating, but it has also been seized by man. He can now reach any part of the sea. He swims and dives and sails and flies. Under his relentless demand for food, cod and haddock and herring have been depleted, even decimated in some areas. Whales, hunted ruthlessly in every sea in which they range, live in small fractions of their ancient numbers, and some are near extinction. Seals are slaughtered in millions. The sea was thought to be an inexhaustible resource only twenty years ago, but now it has come to symbolize the fate of humanity. It mothered man, but it may, if mistreated further, destroy him. If its temperature, or its chemical composition, or its capacity to produce the ubiquitous plankton is upset, it may cease to mother anything.

The balance of the sea's power is contained in those chemical elements moving freely everywhere in the world. The oxygen, calcium, potassium, and sodium are our modern links with the origin of life in the sea, and their fates must be of prime concern to all of us. They formed the first flow of life through the primitive organs of the world's first creatures; it is no accident that they flow today in the veins of whales and lions and men in about the same proportions as they are present in sea water.

13. The expansive ocean encompasses great differences in saltiness, temperature, depth, turbulence. It may swarm with life in places but be relatively barren in others; it is a vast repository of diverse life forms.

14. A larva of the jellyfish Ephyra *floats in Bermudian night waters. This is one tiny member of the plankton world, which swarms in unimaginable numbers at the surface of most seas and is part of the base of the oceanic food pyramid.*

16. A shrimp swims to the surface of a channel in the Great Barrier Reef of Australia. It will join the billions of plankton among which it will hunt and be hunted. At dawn, it will retreat from the surface, swimming downward with the other free-swimming animals of the plankton. When it reaches a safe depth, it will await the coming of its active night.

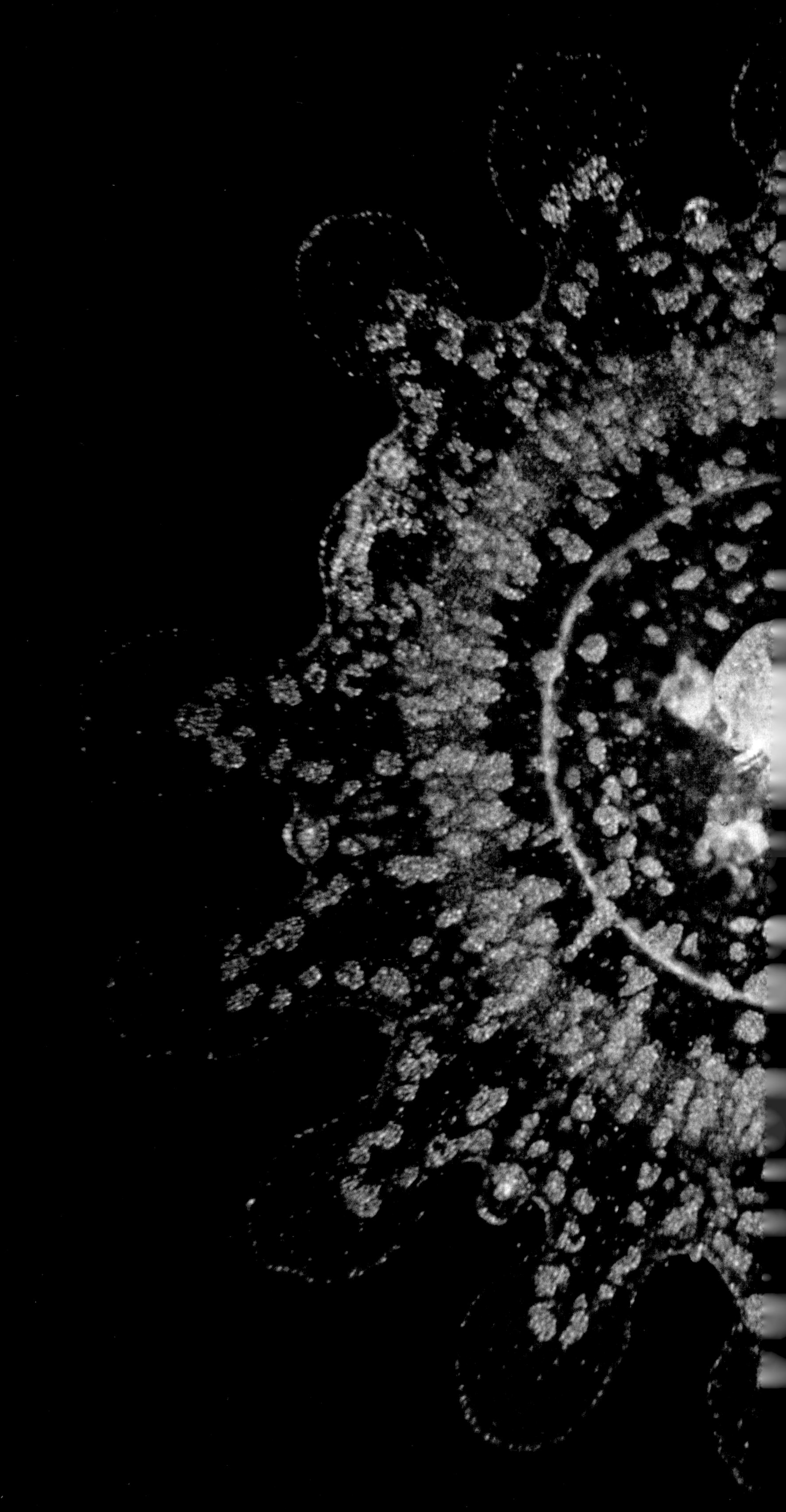

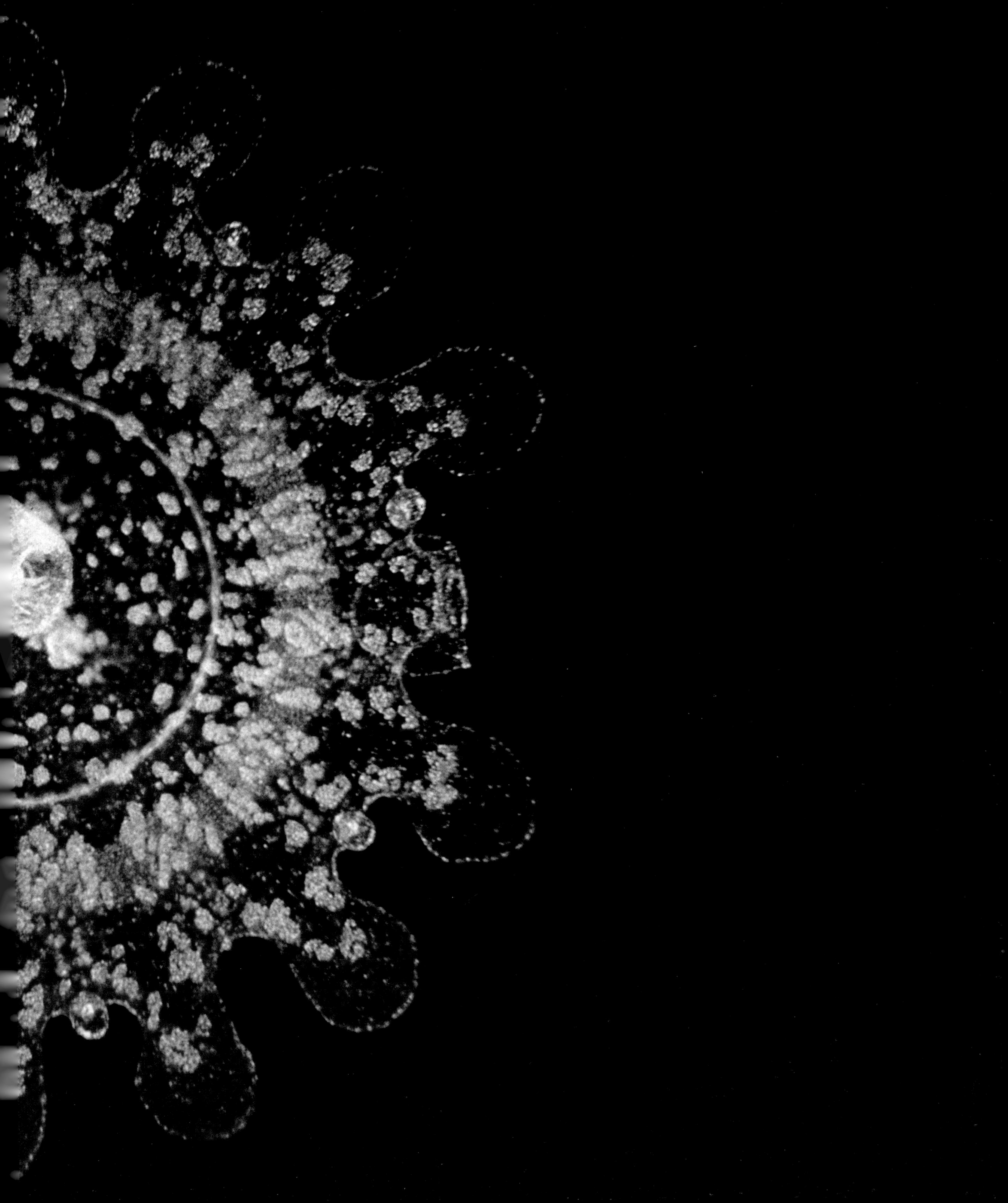

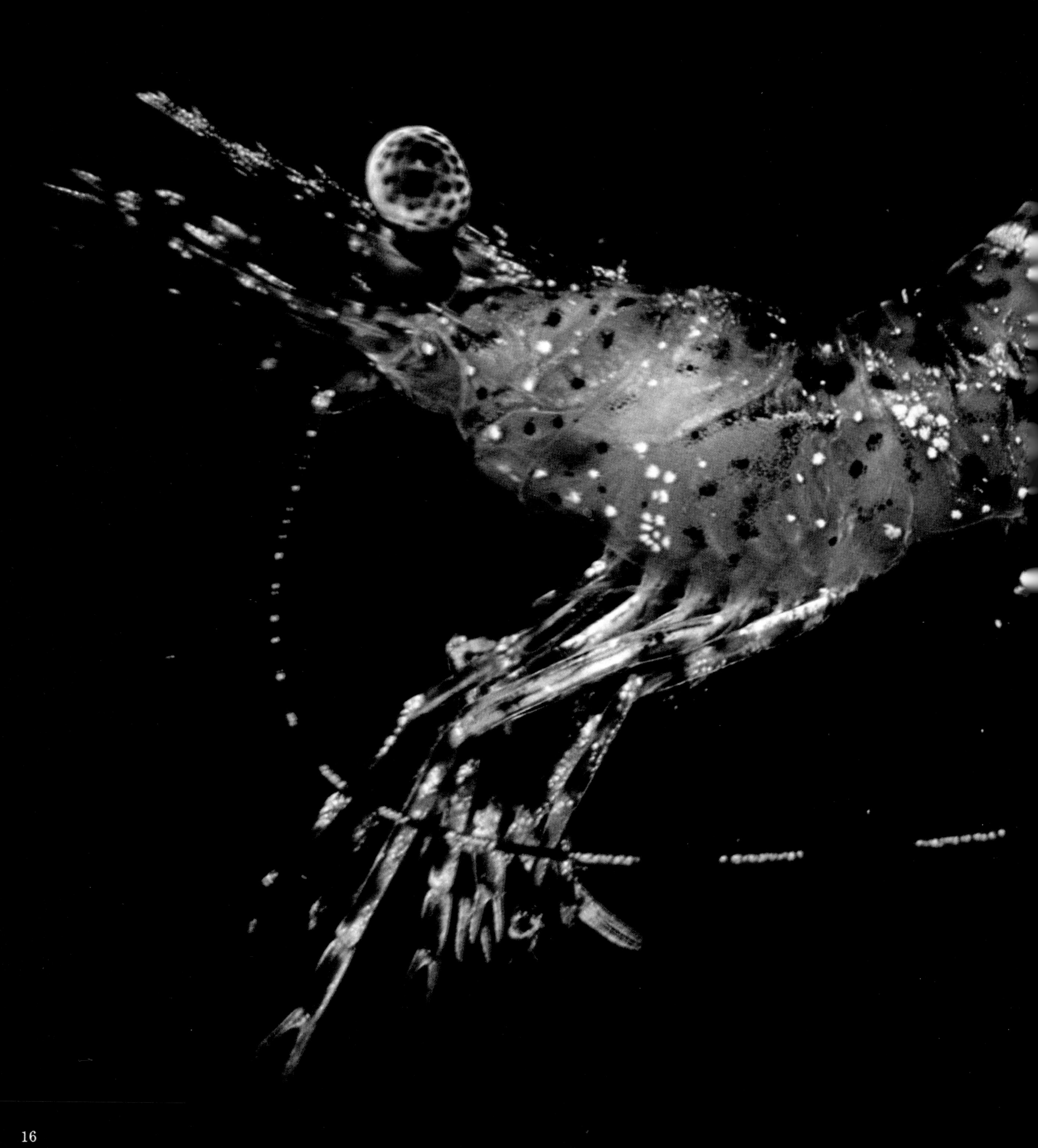

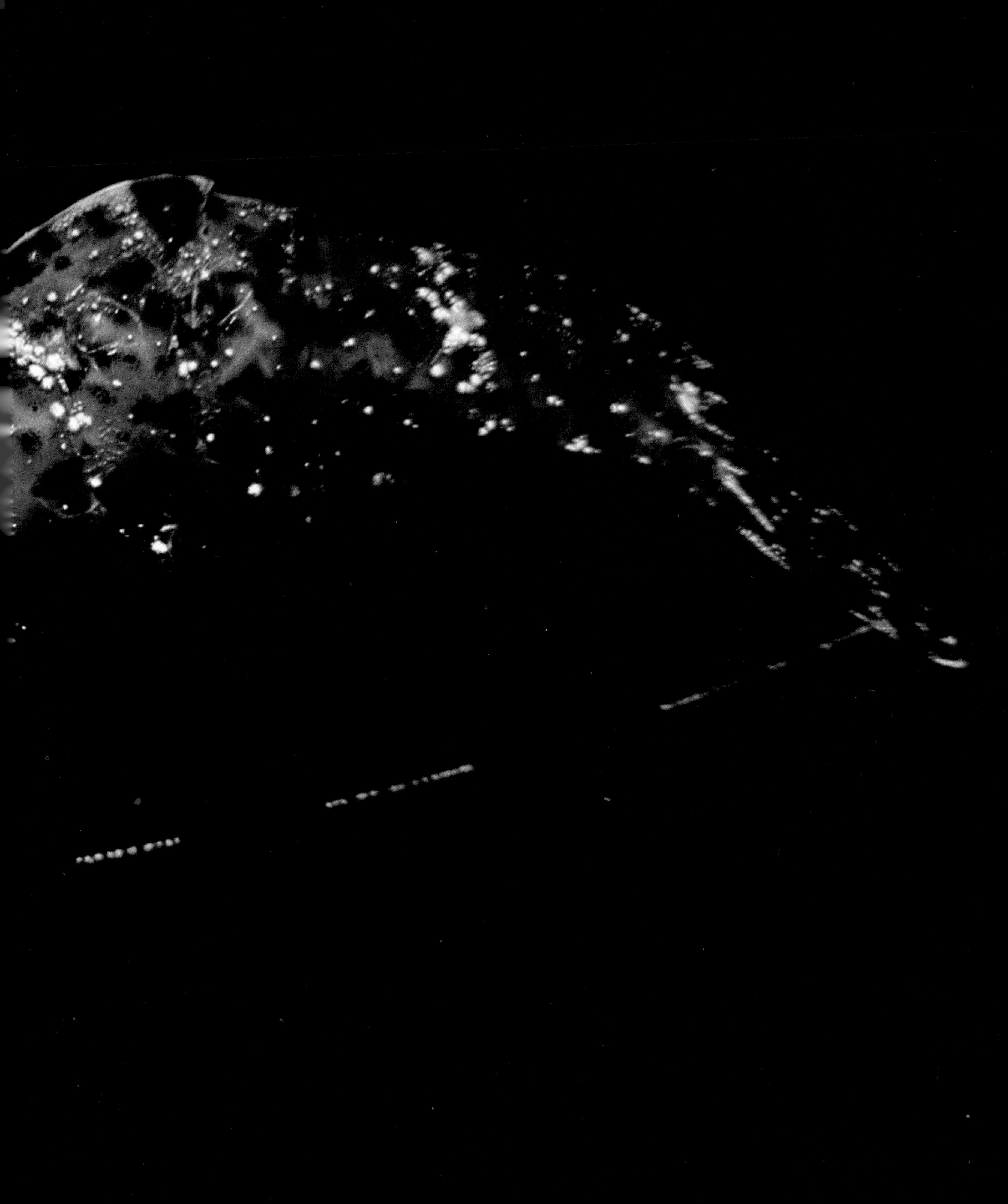

18. A flatworm glides gracefully along under dead coral in the Great Barrier Reef of Australia, its rich colors touched by tropical sun streaming through the clear water. Flatworms, which may be carnivores, vegetarians, or scavengers, are common in all the oceans of the world and can regenerate any part of their bodies.
19. Some large clams in the Tridacna *genus may eventually grow to be four feet long and 500 pounds in weight. In its permanent home on the sea bottom near a coral reef, or in a shallow lagoon, a clam periodically opens and closes its shells as it filters water for its food. Algae, which live in the mantle tissue of the clam, show the bright blues and greens of their massed bodies. No two clam mantles have quite the same hue.*

20. A richly colored and predatory sun star, searching for shellfish, moves through the cold polar waters near Amchitka Island in the Aleutians. Nearby, an onion anemone, fixed to the bottom, waits for small creatures to swim near its outstretched tentacles.
22. To survive at sea demands very special mechanisms of protection. This porcupine fish has many enemies, but it has evolved a distinctive defense. Ordinarily it is a small, flat fish, with sharp spines held close to its side. But it can, when alarmed, inflate rapidly by swallowing air or water. This forces the spines out from its body and offers a formidable defense against predators.
24. A shellfish tries to protect itself with the tough fortress of its shell, but the harlequin tusk wrasse, a brightly colored fish of Australia's coral reefs, uses its sharp teeth to wrench open the shells.
26. The black-tipped shark, cruising the waters off the Pacific island of Palau, is a member of one of the most fearsome groups of creatures anywhere in the sea. Its torpedolike grace displays the perfection of its adaptation to underwater hunting. Its form has changed little for 350 million years.

28. The dusky dolphin roams the temperate southern seas in schools of up to one hundred creatures. This eight-foot-long animal is inquisitive and often paces ships in its coastal waters habitat by swimming close to their bows.

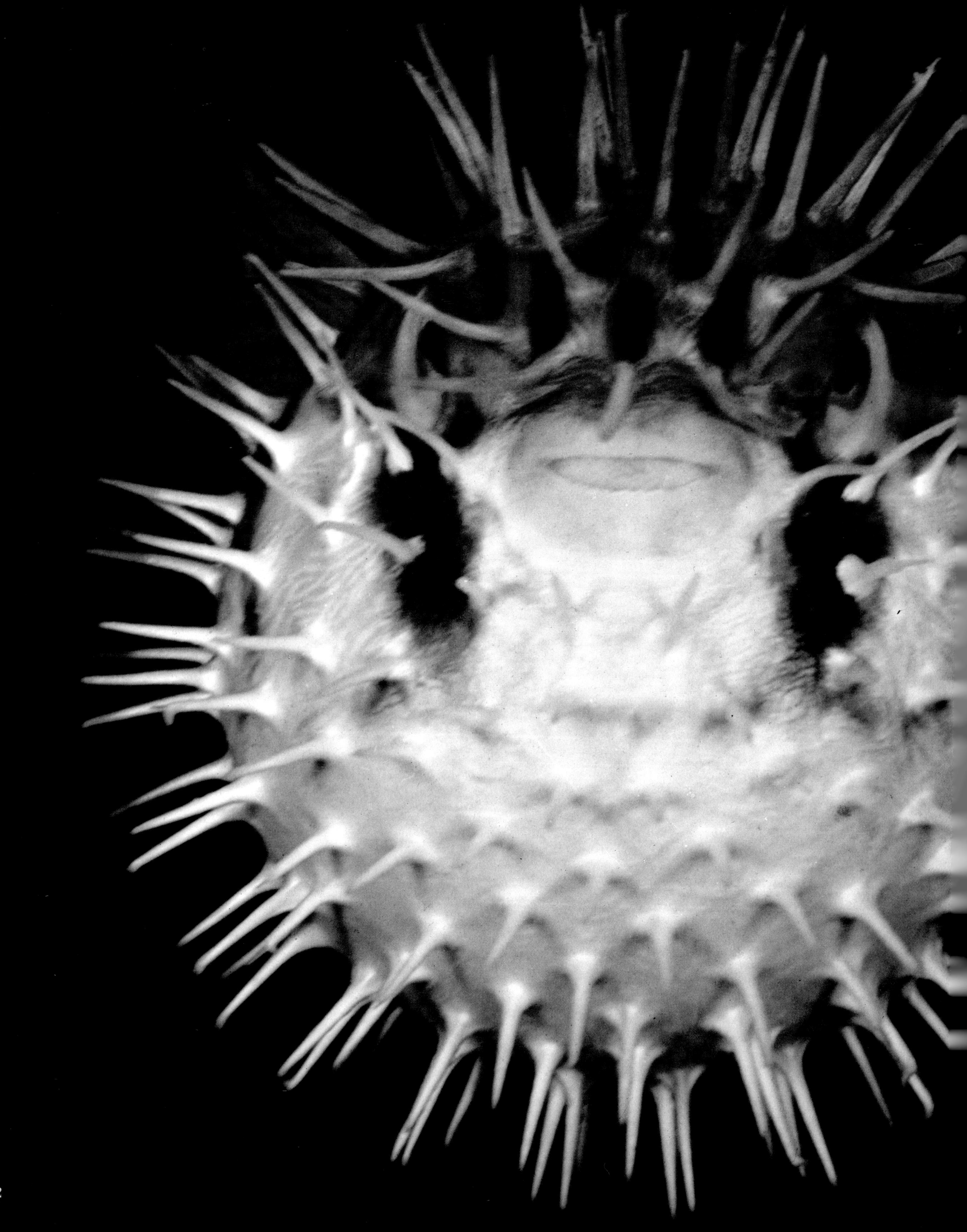

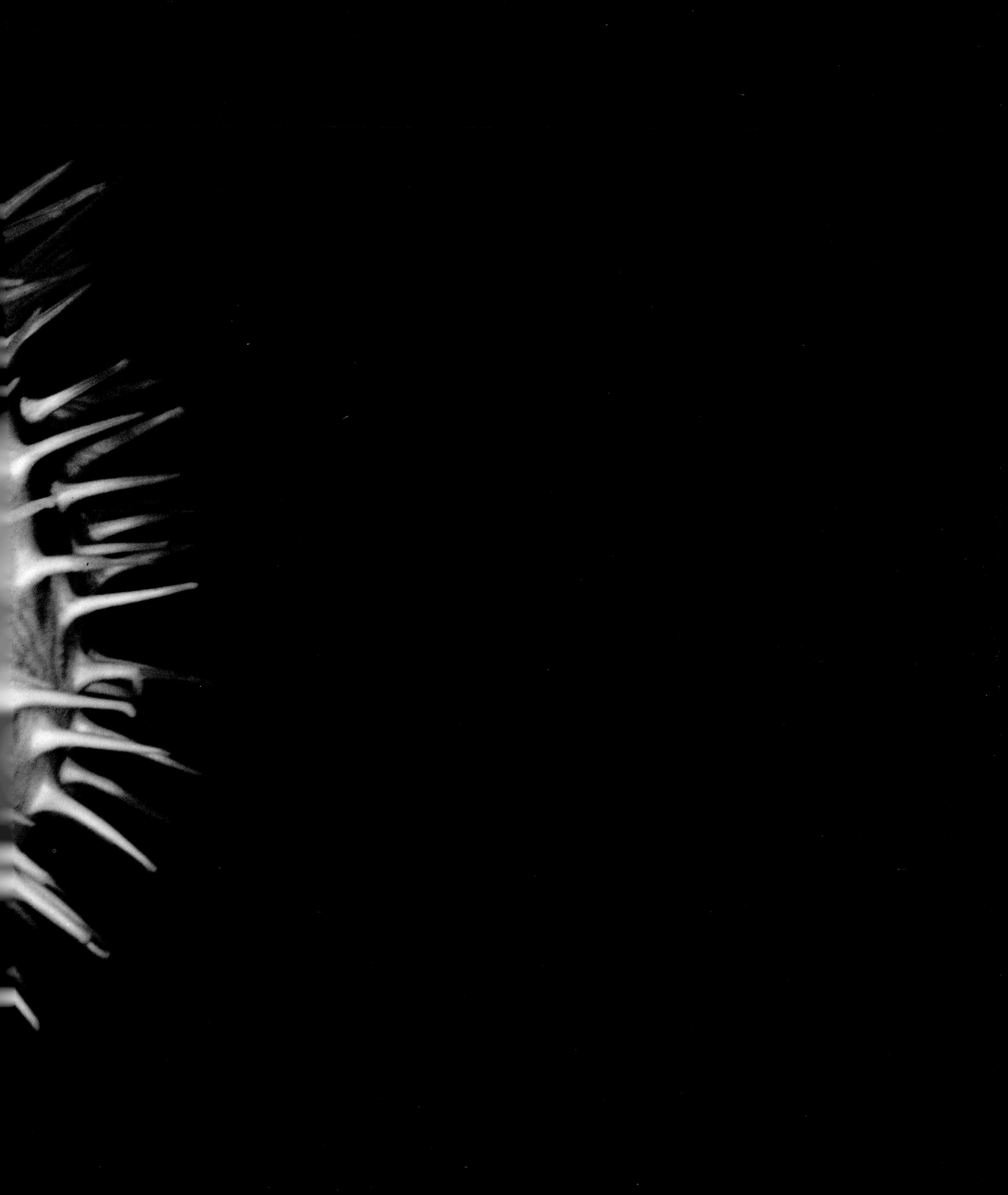

2 Shores of Transition

Photographs by:
Philip Hyde, 33
Dennis Brokaw, 34
Lorus and Margery Milne, 36
Edward R. Degginger, 37
Jessie O'Connell Gibbs, 38
Tui De Roy, 40
George Holton, 41, 43
David Cavagnaro, 42, 46
George Silk: Time/Life Picture Agency, 44
Fritz Goro: Time/Life Picture Agency, 48

It was midnight on the eastern shores of Newfoundland; a bright moon cut a long track across the water from Greenland. Shoals of tiny, herringlike fish were coming toward the beach on which I was standing. These tiny capelin, like some other creatures, had become completely dependent on the shifting sands to bury their millions of eggs and milt. They came ashore to spawn in serried lines, like old-time infantry advancing, and their passage was accompanied by an eerie offshore sound, like the smack of distant clapping. Out there, in the moonlight, the big fish had gathered to eat the massed millions of capelin, and the dusky sea echoed to the steady slap of the hunters' bodies.

But the tiny capelin were determined on survival in the face of death, and they kept coming ashore to deposit their eggs. They linger in my memory; their small bodies pressing against the dark shore gave me a hint of the tension that exists along every one of the earth's coastlines. These frontiers are eternally under attack from the sea; waves smash into the land, reduce its capes and try to fill in its bays, and reveal in the struggle the richly diverse enigma of all life. Many shore watchers have felt the force of universal mystery that comes to them as they stand facing the sea. And, indeed, it is a wonderful varied and magic world: the flashing lights of billions of diatoms, the rotting carcasses of beached fish, the crashing of great Biscay waves, the splash of jumping stingray bodies, the clicking of shrimp, the hurried flight of fiddler crabs, all evoke a picture of life in transition. The sea creatures are held back, for the moment, by the reluctant land, but many of them are bound, in the end, to advance and begin the journey inland.

With its deadly pursuits and unlikely refuges, the shore is a microcosm of the struggle between the hunter and the hunted. Snails and anemones cling together in the curve of abraded shoreline rocks. There is water to sustain them in a small rock pool left from high tides. There is protection for them from the waves that will come with the rising of the tide.

For me, the tenacity of the barnacle and the limpet best exemplify the command to survive in this dangerous world. The conical forms of both creatures are perfectly designed to deflect the immense force of inrushing waves. The barnacle cements itself to rock so tightly that it would be smashed before it could be moved. The limpet uses a suction cup attached firmly to the rocks as its defense against the ceaseless waves. Mussels bind themselves to the rocks with threads, and cluster together by the thousands in a cooperative effort to break the power of the waves. Green urchins splay their thin, hollow feet against the rocks. Each foot is equipped with its own suction disk, and despite its tiny size, the urchin can survive the pounding of the waves.

When I walk the beaches and estuaries of the shoreline, I am always impressed by the ingenuity of its creatures. Worms tunnel to safety in estuarine mud. Whelks, clams, and crabs burrow into sand—some of them so swiftly that I can barely catch them with my spade. Some shellfish eat their way into rock sanctuaries, or into the bodies of other creatures.

Inside this sand, this mud, this shingle beach, is a great, damp, dark world that shelters billions of lives. Here, there is protection from most fish, from the long, probing bills of curlews, oystercatchers, and sandpipers. Here, there is some hope of escaping the stingray, which uses its flat body to disturb the surface of the sand, some relief from the deep-reaching tentacles of the octopus.

The sand is both refuge and hunting ground, the perfect insulator, ideal for the sand dollar working its way on minuscule legs to drive its sharp shell edge downward. It is perfect for the mobile and aggressive crab, which can dig to find food, or tunnel to find safety. It is a haven for countless oceanic creatures. Turtles rise from the shoreline waves, dig their eggs into the sand, and depart for a year. Seals and walruses and sea elephants and other creatures make the sandy shore a place of procreation and relaxation. Watching these creatures confirms the relationship of all life with the sea.

When the tide is low along New England's northern shores, the history of our emergence from the sea is written on the dark, vertical slopes of rock. There, the blue-green algae, which are certainly among the oldest forms of life on earth, live at the highest tide levels. But they still depend on the twice-a-day arrival of the high tides for sea water, and there they remain, a black presence in the pale shore light. Beneath them, snails work along the cracks in rocks, seeking out plants on which to graze. Masses of barnacles are clustered under the snails just below the high-tide line. In some places they are joined by mussels. The long brown limbs of the kelps float, or are stranded, at the lowest tides. Each layer of life tells us part of a story of struggle and evolution.

There are creatures which have gone beyond the high-tide limit to reach into the splash zone. There, only a meager gift of moisture comes in the spray of waves. Some snails of the periwinkle family live in this half-world, and are so well adapted that they can do without sea water for months. The European rock periwinkle has climbed so far that it can only cast its eggs into the sea at spring tides; the rest of its life is quite independent of the sea.

Every time I watch land snails, I am reminded that all of them originally came from the sea, and that all had to cross the transition shoreline to reach inland. The process is continuing; periwinkles are in many different stages of development. The smooth periwinkle is utterly dependent on sea water and hides in seaweeds at low tides. The common periwinkle, on the other hand, may be touched by the sea only at high tides, when it can drop its eggs. The rough periwinkle is almost entirely free from the sea. Unlike the other periwinkles, it has a gill cavity which is advanced enough to be almost a lung, and it can survive more than thirty days without water.

But the success of the transitional shore's animals and plants depends, finally, on the work of the tides and currents. A tide brings water high up the shore at predictable times, supplying a flood of food for the waiting multitudes. Then it withdraws, taking with it a cargo of chemical signals, excrement, released sperm, expelled eggs, and newborn creatures. In this steady movement of water is contained the ultimate message of the shoreline: All living things are totally interdependent. The chemicals of plants, in some mysterious way, affect the behavior of animals. Spawning creatures send signals far along the shore and arouse other creatures. Nothing is sacred unto itself. Everything is connected to everything else. This is a verity we ignore at our peril.

Despite the complexity of coastal life, there are only four types of shoreline—sandy, rocky, muddy, and coralline. Sand is generally born on the shore, the result of the relentless erosion of seashore rocks by the crushing waves, though it can be manufactured elsewhere and brought to the shore by rivers. It is mostly made of crushed quartz, and is a residue of the earth's violent history, the product of its constant erosion. Black sand is ground volcanic rock; soft, almost chalky white sand is crushed coral and powdered shells; white, gritty sand is pulverized mountains. Yet diversity rules here too: Sand may be black, white, yellow, greenish, lavender, or red.

The coral shore is formed by a community of animals cooperating to resist the power of the sea. The Great Barrier Reef, off the coast of Queensland in northeastern Australia, is probably the most beautiful coral shoreline anywhere in the world. It is a triangular-shaped series of reefs running for more than twelve hundred miles along the coast, and is a spacious refuge for shoreline life. When I went skin-diving among the reefs, I saw the endlessly fascinating interchange that goes on among the marine creatures living there.

The reef builders are tiny animals which continuously grow calcareous skeletons. They live where the water seldom drops below the 70-degree mark, and they have only been able to settle off Queensland and in the Pacific, on the African east coast, and along the shores of Brazil, the West Indies, and southern Florida.

I knew that the little animals were quiescent, as I swam slowly through their multihued colonies. They would burst from their protective cups to feed that night. Their extruding tentacles, equipped with stinging cells, would wave blindly in search of passing plankton. I examined sulphur sponges, which are able to dissolve the coral with special chemicals. These creatures serve as a benign check on the growth of the uninhibited reef builders. I swam by hidden shellfish which had honeycombed the reefs with tunnels. They were neighbors of the many worms that also chew into the coral for shelter. Another reef dweller is the female gall crab. She digs a small cavity in the coral and uses it as a home. But once she is fully grown, she is likely to be imprisoned in her refuge for the rest of her life. Thus held, she draws sea water and strains out its food. She receives a tiny, free-swimming male visitor when her time comes to breed.

So great is the bulk of calcareous material that the corals appear to be inviolate. They have built millions of tons of the rocklike substance, yet nothing along any coast is truly secure. The crown of thorns, a type of starfish, has unexpectedly become ascendant in Queensland and is killing the corals in such numbers that the Great Barrier Reef is dangerously threatened.

Danger is implicit along any shoreline. During warm water spells, Mediterranean octopuses occasionally are able to reach the southern shores of Great Britain, where they eat everything they can catch before the cold water returns and wipes them out. Beach sand is drilled constantly by whelks to reach buried shellfish. Even the billions of barnacles are not safe on their rocks. When hordes of dog whelks arrive, they can decimate the barnacles in one year. Mussels move quickly into vacant barnacle territory, but they are eaten by the whelks too. Only when their supplies of food are gone, and they are starving, do the whelks disappear, making way for the barnacles to take over once again. Nothing is certain at the shore, except the imperative that for each creature which dies, another will fill its place.

Although shoreline dwellers appear to be simpler creatures than the larger, terrestrial forms, their ingenuity suggests that man may not really understand the complexity of the life force. The razor clam projects its foot half out of its body, expanding it to form an anchor. Then a long muscle rapidly withdraws the foot, so that the clam is pulled deeper into the sand. The green crab uses special muscles to snap off the broken part of a limb, then closes the wound with a coagulant to stop the flow of its colorless blood. The large queen conch jerks its massive body in a series of extraordinary leaps that tumble the massive shell forward.

The tension of the shore zone forces life to plumb its limits. And the shores of islands offer creatures unique opportunities to test themselves. Lizards use the rocky coasts of the Galapagos Islands as vaulting boards from which to return to the sea. Watching them is like witnessing a re-enactment of ancient times, when some mammals returned to the sea to be transformed into whales and porpoises. The bare volcanic rocks of the Galapagos shorelines are now the lizards' one narrow contact with the land, as they regularly go to the sea to graze on offshore beds of seaweed.

Sea birds and shore birds use the shore in similar ways. For some, such as the ocean-ranging gannets and boobies, murres and puffins, the shore is a place of brief refuge where they can breed. For others, such as frigate birds, pelicans, and gulls, it is a long and narrow band of hunting territory which satisfies all their needs.

The tension existing along the transition shore is so extreme that every weapon, every possible skill, is thrown into the life struggle. Here, life is visibly tough and dangerous. The real tension of life is displayed as it crosses the strand. The push and pull of the tides, the great force of the waves, the sudden warming or cooling of the waters, the variety of shelters, all combine to make a perfect meeting place for the chemicals that the land has given the sea. The creatures of the shore are the living expression of that miracle.

33. This California beach represents the ever-changing transition between the land and the sea. To survive here, shoreline creatures must be able to tolerate a continual alternation of wet and dry conditions.

34. Life on the seashore demands common qualities from different species of creatures. Sea anemones and turban snails cluster together at low tide in a tidal pool of the rocky California coast. The tiny world of the shallow tide pool is sustained only by the small amount of sea water left in it when the tide retreats.

36. The body of this keyhole sand dollar is flat and covered with fine moveable brown spines, which give it powerful traction to drive its body out of sight when an enemy threatens on the sandy bottom of the Atlantic coast off North America.
37. A ghost crab, poised on a California beach, has made a special adjustment to life on land. But it must return occasionally to the sea to fill its gill chambers with sea water. Its link to the sea is expressed even more strongly when its young grow to maturity as members of the drifting plankton community. Then they return to the shore to live as adults.

38. An Atlantic loggerhead turtle returns to the ocean after laying her eggs on a beach on the southeast coast of the United States. She has spent most of her life at sea, coming ashore only to deposit her eggs in the sand of her ancestral birthplace. Her young will make their way back to the ocean immediately after they hatch from their damp sandy nest.

40, 41. The rocky shores of the Galapagos Islands are home for the extraordinary marine iguanas. They bask for long hours waiting for low tide. Then they go into the sea and swim to the bottom to feed on marine vegetation. Although their kind were once entirely dry land creatures, they can stay submerged up to one hour. Their specialist shoreline life is emulated by the small orange crab, the sally lightfoot, which is found on cliffs overhanging tropical shores in many parts of the world; but only on the Galapagos do such crabs pick ticks from beneath the skin of marine iguanas.

42. A brown booby hovers over the transparent waters of the Pacific, a master of graceful flight but able to plunge instantly after a fish underwater. It feeds its young in nests on Pearl and Hermes Reef in the Hawaiian archipelago.
43. On land, boobies are clumsy, almost helpless, quite unlike the graceful and powerful birds they are in the air. Here a blue-faced booby feeds its chick on Hood Island in the Galapagos. It furiously defends its nest against intruders. These boobies nest in loose colonies by the sea where each bird cares for a single offspring.
44. Every shore provides its distinctive conditions on which life thrives; these white pelicans hunt along the sheltered waters off the southwest coast of Africa. Once they find a school of fish, they will gather in loose formation, swoop down low, and then splash wings and feet to drive the fish into the shallows, where they can feed on them easily.
46. A monk seal, its pup bred in a colony on a reef in the Hawaiian archipelago, guards and suckles its young until it is weaned and ready to begin its independent life in the blue waters of the central Pacific.
48. Great walrus tusks, glowing in the late afternoon Alaskan sun, are weapons that fit these bulky creatures for their fierce territorial disputes on small islands during the mating season. The tusks are also used for gouging up shellfish deep underwater.

3 The Haunted Antarctic

Photographs by:
Emil Schulthess: Black Star, 53
S. D. MacDonald, 54
George Holton, 56, 58, 60

The earliest visitors to Antarctica were puzzled to find that the continent contained petrified tree trunks, exotic leaves imprinted on rocks, and coal seams jutting from the frozen ground. One 1912 visitor, Charles Hedley, imagined that in a more temperate past there had been a garden "of rippling brooks, of singing birds, of blossoming flowers, and forest glades" in the heart of Antarctica. That this great glacial land mass once teemed with animals and plants is true enough, but the notion of rippling brooks and blossoming flowers is imagery best suited for poets. I can find more marvels crossing Antarctica's icy vastness today than in contemplating its real or imaginary past.

It is the continent of the widest extremes. It has the strongest winds, the thickest ice, the lowest temperatures, and the greatest isolation. It is a desert. Rain drops only in tiny territories, and annual snowfalls are measured in inches. But its ice is a mile thick. A volcano steams on its shores. It has giant interior mountain ranges, and one of those peaks, Fridtjof Nansen, which is 19,000 feet high, has never been scaled. By ordinary human standards, almost all of the Antarctic is inaccessible.

For me, its hushed, piercingly cold valleys, mountains, and ice packs have an eerie, almost haunting quality. I have listened in vain for the cry of a bird, or the crackle of melting ice. But there is only stillness. Walking through the pervasive silence of the Antarctic desert, a man may find the mummified body of a seal twenty-five hundred years old.

Despite the harshness of its welcome, the Antarctic is ringed with immense populations of creatures. These are animals which have found many ways to beat the cold, fight the wind, deny the drought and desolation. Their invincible capacity to overcome these forces is a reminder of our own vulnerability. The Antarctic suggests, more than any other part of the world, that we are only transients on this planet.

The emperor penguins, four feet tall and up to ninety pounds in weight, look oddly humanoid in their immaculate black and white plumage as they come out of the Antarctic Ocean each autumn. They walk scores of miles across the ice to reach their breeding grounds. They are the only birds on earth, and one of the few creatures anywhere, that court and mate while blizzards scream and the temperature goes down to 85 degrees below zero. The female emperors lay their eggs and return to the sea. The males take the eggs on their feet, cradling them from the ice, and fluff their breast feathers forward to conceal them. The wind whips ghostly wraiths of snow among the penguins, giving them the look of round-shouldered children as they rock in unison against its force. The emperors stand there for two months without food until the eggs are ready to hatch and their mates come from the sea to relieve them.

Another creature, the Weddell seal, also faces the winter without retreating. A pack ice animal, the seal swings its head violently to drive chisel-like teeth into the underside of the pack ice until the teeth cut through to the life-giving air above. Unlike the emperor penguin, however, the Weddell seal would quickly die in the hurricane blowing above it. After a full draught of air, the seal dives deep into the pitch blackness of the sea in search of squid. Such is its adjustment to the Antarctic that it can remain underwater for an hour without surfacing to breathe, and can reach depths of nearly two thousand feet. Both seal and penguin battle the wind in their special ways, but it is the killer wind that, indirectly, helps make their lives possible. It sweeps down from the mile-high slopes of the continent, howls across the ice pack, and out to sea. There, it collides with warmer air and is twisted by the Coriolis force of the rotating earth so that it turns eastward to move with a vast current—the Antarctic Convergence—which endlessly circles the continent. Within the convergence, the icy, oxygen-rich waters of Antarctica meet the warmer, mineral-salt-laden waters of the north. Salts, oxygen, cold, and warmth come together in a turbulent union to create a rich mix of oceanic life spread over 12 million square miles.

To describe the convergence as a current is an oversimplification. It bends, twists, expands, contracts. Other currents race under it. Countercurrents spin away from it. Its plankton and krill have adjusted their life cycles to breed in its flow, though some creatures ride surface currents south from the convergence to reach Antarctic waters where they breed. Others drift east at the surface to catch westward-flowing submarine currents that will return them to their birthplaces.

The convergence, directly or indirectly, is the genesis point for a host of lives, and also the demarkation line that separates the Antarctic from the Subantarctic. It is patroled by albatrosses of many species, shearwaters, and countless petrels. The five species of prions—dove-sized petrels—are so numerous that no attempt has been made to count them. One bird watcher, after sailing through endless flocks of them, and knowing that they encircled Antarctica, observed that they must live in "the thousands of millions." The convergence draws creatures to it from both the north and the south, or, from its creative waters, spins away a cornucopia of food that reaches both the far south and the more temperate waters of the north. Its richness of life enables many seals, and sea lions, and penguins, to inhabit the islands of the deep southern hemisphere, animals which, however, do not cross the convergence to hunt in the Antarctic.

Before whaling days, the convergence's unbelievably rich production of plankton, and the practically inexhaustible populations of shrimplike red krill, brought armies of whales into these waters. The baleen whales filtered tons of these creatures from the sea through rows of flexible bones hanging from their upper jaws. The giant blue whale, using its baleen bones as a sieve, needs about three tons of krill every day to sustain its 150-ton bulk. Multitudes of small crabs and fish pursue the two-inch-long krill, while bigger fish and squid hunt the krill eaters. They, in turn, become food for the larger animals, the penguins in their millions, the deadly killer whales, the legions of seals, and the echelons of sea birds fluttering at the surface.

The generous mix and spread of the convergence's life makes possible the great number of crabeater seals, Antarctica's most common large animal. More than ten million of these creatures are spread around the continent. To capitalize on the krill, the crabeater has developed teeth that perform the same function as the bony plates in the mouths of baleen whales. But instead of filtering the krill through the baleen, the crabeater sucks in a mouthful of krill-filled water, then forces it out through its interlocking teeth, effectively trapping the krill.

Unlike the crabeater, the rare Ross seal has curved and spiked teeth adapted to hunting for squid that swim in deep waters. Although the squid are numerous enough, they nowhere match the overwhelming krill populations, which is why there are fewer than fifty thousand Ross seals.

Nearer shore, the most populous of Antarctica's penguins, the Adélies, winter on the ice pack. They feed on fish and squid, but their hunting can be dangerous since the voracious leopard seals and killer whales are in the water waiting to attack them. But a threatened Adélie can fling itself seven feet out of the water to the safety of the ice. In the spring, the Adélies move inland to breed all around Antarctica in huge rookeries. They use pebbles as nesting material and are ceaselessly harassed by skuas—gull-like birds which behave in the predatory fashion of hawks, eating the penguin's eggs.

None of the other penguin species have been able to match the hardy adaptation of the emperors and Adélies, and many of them breed on subantarctic islands that ring the continent north of the convergence.

One other bird, the snowy petrel, has made the Antarctic its permanent home. Unlike other petrels, it never leaves the far south and does not prey on krill as do others of its kind. Instead, it haunts the fringes of the ice pack and hunts small fish swimming near the surface.

To survive in the great cold of Antarctica, where summer temperatures rise only a few degrees above freezing, its creatures have developed an exceptional resistance to cold. Fat, or flesh impregnated with oil, are both good insulators, particularly when combined with thick fur. In addition, seals shut off the blood supply to their skin so that

53. In the world of Antarctic ice, the shifting patterns of moving ice floes give seals, penguins, and sea birds entrances and exits for hunting in the chill dark waters beneath. Here, nearly all hunting, resting, and mating is done on and around the ice. Only small parts of the great continent's land mass are free of ice.

while the surface of their body may be a degree or two above freezing, their internal temperature is maintained at 100 degrees.

All the Antarctic birds have two temperature systems; one allows their webbed feet to reach nearly the freezing point, while the other keeps their bodies warm. Only the pure white sheathbill, which resembles a pigeon, manages in Antarctica without webbed feet. It survives by scavenging around the colonies of other birds and eating their eggs.

The Antarctic Peninsula, a long and narrow promontory where ice does not dominate the land, juts out from the continent's roughly circular mass. From its tip to the point of Tierra del Fuego, the southernmost part of South America, the distance is a little less than one thousand miles, but for most migrants it is an impassable gulf.

Only two flowering plants have reached the Antarctic Peninsula, which has the mildest climate in Antarctica. Sparse grasses, tenacious lichens, and mosses cling to the unyielding land, yet these pioneers are enough to give sustenance and shelter to more than forty species of ticks, mites, lice, springtails, and a peculiar wingless fly.

In the freshwater ponds lining the continent's coasts I have witnessed the irresistible capacity of creatures to find homes in improbable places. Many of these ponds teem with single-celled protozoans and with rotifers, water bears, flatworms, threadworms, and other tiny creatures and plants. The connection between these isolated life forms and the great Antarctic Convergence, hundreds of miles away, becomes clear when rich nutrients from the sea creatures drift into the ponds on strong winds. The plankton and the krill finally reach and sustain Antarctica's few land creatures.

But from the continent's shores, the march inland grows less certain, then peters out. Parasites and other small forms of life can reach the coast in the plumage of birds or on the larger sea creatures, but they cannot face the terrible overland journey. Only the lichens and a few bacteria have managed to colonize some of the bleak rock slopes of inland mountains, which rise above the thick ice, because they were able to travel as scraps of matter in the grip of Antarctic winds.

I first flew to the South Pole many years ago, but that flight was premature. I knew nothing then of Antarctic life. I understood only vaguely that Antarctic wind affects much of the southern hemisphere's weather, and that Antarctic waters, reaching far to the north, influence most of the world's oceans, and therefore modify climates everywhere. I was an ignorant young man and did not understand that one day I would be made to feel insignificant by winter-nesting penguins, wingless flies, and seals that eat plankton. These strange creatures did not have special meaning for me until I began to understand something of the quality of our own struggle for survival on this earth.

54. A superb adaptation to a world of ice is exhibited by the winter-breeding emperor penguin. The ninety-pound, four-foot-tall males incubate the eggs by balancing them between the tops of their feet and the lower parts of their bodies.

56. Royal penguins scramble through shallow water to reach their inland rookery on lonely Macquarie Island, southwest of New Zealand. There, in a colony numbering two million or more, they will feed their young with regurgitated fish.

58. Each creature makes its own unique adjustment to the shelter that it finds available. Most seals and sea lions make uneasy landfalls on open beaches. But this Hooker's sea lion stands in a grove of rata trees on Enderby Island, one of the subantarctic Auckland Islands.

60. True Antarctic creatures, crab-eater seals sprawl on an ice floe. These beautiful seals encircle Antarctica in the tens of millions. They are misnamed; their teeth are adapted only for filtering krill from plankton-rich waters.

4 The Primeval Wetlands

Photographs by:
Gordon S. Smith, 65
Larry West, 66, 68
Jack Dermid, 70
Patricia Caulfield, 72
Thase Daniel, 73, 74
M. Philip Kahl, 76

The Iowa sky was intensely blue and patched with rounded white clouds as I paddled between two islands packed with cattails. Ahead of me, an expanse of unruffled water, almost completely surrounded by vegetation, posed a truly primeval picture in the expanded vision of my powerful binoculars. Large dragonflies darted low over the water. The protuberant eyes of a score of frogs broke through the surface among the flat leaves of water lilies. Insects zigzagged in the calm air.

One dragonfly whisked to the stem of a sedge and stood there, its trembling wings shimmering in the sun. The dragonfly, its kind unchanged in form for many millions of years, gave me a vivid visual link between the shrinking wetlands of today and the great bogs, marshes, and shallow lakes of the ancient past. Here, within reach of my paddle, was an exemplar of the quality of prehistoric life. The insect, having not changed its form, surely had not radically changed its habits.

The dragonfly zipped across the water and flew directly into a small clot of tiny insects—gnats, probably—which were dancing at the surface. The dragonfly lowered its hair-fringed front legs and trapped the gnats in this ingenious net. It continued to dart back and forth across the water as it pushed its proboscis into the net and sucked dry the bodies of its victims. I was observing the rapacious behavior of a perfect hunting creature. None was more efficient or more deadly than this beautiful insect. Its great compound eyes could see in every direction. Its head was jointed so that it could look under its body, or behind it. It could fly sideways or backward. It was a cannibal with catholic tastes—mosquitoes, gnats, butterflies, wasps, flies, and smaller dragonflies were killed with such dispatch that it appeared mechanical, perfect. Beneath the water lurked the dragonfly larva, one of the most ferocious of all underwater creatures, and powerful enough to eat tadpoles and small fish. The dragonfly was a product of the water, and would return its eggs there when the time was right.

The primeval quality of this sheltered patch of Iowa marsh water was enhanced when I looked down and saw many tiny lives darting and drifting in the shallows. They, too, had changed little over millions of years in the life-giving, life-sustaining water. The phosphates, calcium carbonate, nitrates, and oxygen made it a veritable soup-mix of creatures. Microscopic diatoms and desmids were reproducing apace; algae swarmed along the stems of every plant; immeasurable numbers of aquatic insects in every conceivable shape and stage of development provided food for the bullheads and minnows which thrived in deeper water. Grebes and coots and rails trod the shallows among the muskrats, foraging raccoons, and ferocious snapping turtles.

The wetlands harbor the most primitive and the most complex of creatures. Nowhere else do armies of amphibians live side by side with elegant egrets, herons, storks, and cranes. The water moccasin swims near the bobcat. The gnarled body of the alligator lies sharply distinct from the gorgeous scarlet ibis passing overhead. The wetlands put evolution on display, and that is their fascination; the sweep of biological time is compressed into one picture. While the diatoms and desmids ceaselessly divide, as they have been doing for hundreds of millions of years, ducks and geese come down out of the skies in cacophonous flights that churn the wetland waters into foam.

It is difficult to set the glacial marshes of North America beside images of southern wetlands where ice is a rare visitor. Yet slender links join them. The South Dakota marshes were scoured and filled during periods when ice covered the land. Now, avocets, willets, and other shore birds patrol the muddy banks as ducks pump overhead, and bitterns sound their rich calls in counterpoint to the musical cries of red-winged blackbirds. This is friendly territory to the human watcher, and is in hard contrast to the southern swamps, some of which were scooped out by Atlantic waves, and then enclosed. There, in the spring, the primeval bellowing of alligators rolls through the labyrinthine shadows of giant cypresses.

In such a juxtaposition, it is the reptile which gives me the greater feeling of endurance and eternity. The Dakotan swamp birds somehow seem ephemeral, mere decorations topping the great reptile heritage that lies in the waters beneath them. Though I have seen ospreys carrying wriggling snakes into tree tops in Georgia, which would seem to indicate the supremacy of the bird, I have also watched an alligator slowly drown a gasping raccoon. The beautiful crowned cranes delicately picking their way through the shallows of African wetlands must be measured against the torpedolike lunge from the water of a crocodile, the sweep of its broad tail, and the death of a full-grown zebra. In this wetland meeting of reptile and warm-blood, the evolutionary process combines a mixture of unlike forms.

The wetlands offer each group of creatures opportunities for their special skills. The southern swamps favor the snakes. One scorching hot day, I drove my canoe into cypress shade and came upon a snake. The pupils of its lidless eyes were thin slits of deep green. Frozen for a moment, I watched its mouth open to reveal snow-white flesh; a venomous cottonmouth was warning me to keep my distance. At night, in the summer heat of the southern Great Dismal Swamp, my flashlight picked out racers, king snakes, pine snakes, and indigos slipping along in search of sandy places to lay their eggs. It was a dangerous time to be walking because rattlers and cottonmouths, which bear their young alive, were active also.

But the turtles have made the wetlands their own more than any other kind of creature except the water bird. A snapping turtle, cold hard eyes canted upward, trailed a string of ducklings following their mother from a swamp-side nest, and I was witness to their terror as the snapper rose twice and brought down one of the youngsters in its cruel mouth. Its counterpart in the south is the alligator snapper, a giant which may top 150 pounds. The alligator snapper lies motionless in deepish water, a small pink cylinder of flesh protruding from its mouth. The flesh looks like a worm, and when a fish is fooled by this pseudofood, only a fraction of a second elapses before the fish is seized by the large hooked mouth of the turtle.

Despite their armament, despite their deft swimming through the water weeds, turtles must eventually face their warm-blooded enemies. No turtle has evolved the capacity to hatch its eggs in water, or bear live young, so, inevitably, turtles come out of the water to dig their eggs into sand or earth along wetland shores. Just as inevitably, the bears and raccoons patrol the same territory and dig up the eggs. The turtles become victims of fellow reptiles as well. Snakes, which seem to have an uncanny capacity for knowing where turtle eggs will be laid, often wait nearby while the turtle lays her clutch. When I watched a large king snake rooting with its blunt head into the earth covering freshly buried turtle eggs, and then swallowing eighteen of them, I understood that I was witnessing a very ancient drama in the evolution of life in the wetlands.

Equally compelling in its evocation of ancient geological ages was my sighting, one hot summer day in the Okefenokee Swamp, of a female alligator tearing apart her roughly built nest of rubbish and mud. She had heard the first grunt of a hatched youngster, and soon the young alligators were spilling out of the wreckage of the nest. In the bright sun they looked like pretty, mechanical toys. Sleek as silk, and agile, they sought to swim away while their mother tried to stop them, corral them, until finally they were ready to climb up on her back to sunbathe. It was a scene of parental concern straight from the pages of warm-blooded natural history.

While the amphibians and the reptiles may most accurately call the wetlands, or the marine estuaries, their original homes, it is the birds which have now thoroughly colonized these regions. I remember one early fall evening in a Canadian prairie marsh when thousands of ducks and geese were becoming agitated by the onset of their migration fever. The ducks circled the marsh in quacking groups—mallards and gadwalls, green-winged teals and pintails, shovelers and baldpates. In their excitement they landed and took off like tiny, animated fighter planes, showing the bright blues and greens, the fawns and browns of their plumage in the light of the falling sun. The Canada geese made practice flights across the marsh in V-formation, keeping themselves apart from the snow geese and the brants. Visually, at least, all other life became subordinate to those hordes of birds.

Wading across a southern swamp, I was witness to a totally different conquest of the wetlands by the birds. Egrets stood in cool shadows, preening their beautiful white plumes with exaggerated care. A great blue heron, hunch-winged and skew-necked, deftly snapped a frog up from the shallows. With crashing surprise, a five-foot-high sandhill crane, until now camouflaged against the vegetation, took to the air with bugling calls of alarm. In the distance, ibises marked graceful curves against the pale blue sky as they turned toward their noisy, reeking rookeries, which were set high in rough-stick nests. The uproar signaled the raising of several thousand nestlings. This was the country of the peculiar anhinga, or water turkey, which dives and chases fish underwater with deadly speed, although it does not have waterproof feathers.

Here, for a moment, the birds diminished the hidden reptiles and amphibians, the original forms of life, to mute players in the drama of the wetlands. And when I paused for a moment, I noticed that several hundred wood ibises had formed an upright column in their circling flight. The birds were piled on top of one another with a peculiar and touching orderliness. Apparently, they had discovered an updraft rising from the open waters. While I watched, they circled higher and higher until the updraft's force was dissipated and the ibises scattered and disappeared.

Then, with the day fading fast, as it does in the south, I caught sight of a mass of flying creatures approaching me across the weed-spattered surface of the shallow water. They were dragonflies, all bound for some common destination, and oblivious to my presence. Several hundred of them passed on either side of me, possessed by some purpose I could not know.

When they had disappeared, I was left with a thought. Here, in these primeval wetlands, the ancient reptiles and amphibians had survived and evolved alongside the more adaptable and more accomplished newcomers—the birds and the mammals. This had been made possible by the stability of the environment itself. If the *place* was right, then almost anything was possible in life.

65. Lily pads glimpsed through a screen of reeds are the visible parts of a teeming world of marsh life in which a host of different creatures and plants find refuge.

66. Dew spread like diamonds on the resting dragonfly's wings provides one bright image of the water world of the marsh. The dragonfly begins life as a nymph crawling on the muddy bottom. As an adult, it flies across the marsh hunting flies and mosquitoes.
68. Eye-deep in shallow waters, a green frog waits in motionless suspension for a victim to fly near it. Little changed in form for millions of years, the frog is an efficient aquatic hunter.
70. The eastern cottonmouth, or water moccasin, sunning itself in a North Carolina swamp, has returned to the waters whence reptiles originally came. It now leads a dual existence in water and on land. A venomous and deadly hunter, it is itself hunted by hawks and other predators.

72. A silent, motionless predator, an alligator in the Everglades of southern Florida waits half submerged for an unwary garfish or a careless turtle.
73. Brackish wetlands once harbored numerous redheads, particularly the Louisiana bottom lands. But when the wetlands were drained, this common North American diving duck became rare.

74. A triptych of egrets is displayed in the shallow waters of a Louisiana bayou. These statuesque birds, when motionless, appear to be graceful paintings set against a serene background, common egrets in the two end panels, a snowy egret in the center.
76. At evening, scarlet ibis return to their colony in dense mangroves in a swamp on Trinidad.

5 Spectacular Prairies

Photographs by:
David Plowden, 81
Les Blacklock, 82
Bill Browning, 84, 86
Glenn D. Chambers, 88
Jen and Des Bartlett: Bruce Coleman, Inc., 90

When white men appeared at the edge of the prairies, they recoiled from what seemed to be "an impassable desert." Despite the lushness of the grass, the land was too open for these first white men. It had no shelter, no clustering trees. It stretched out of sight, grass to the horizon. During long droughts, it became so dry that the kick of a sage grouse's foot raised thick dust. When the buffalo moved, the dust obscured the sun. The chinook, in its icy arctic form, had the force of a hurricane. "No," men said, "this is no place for us to live."

Perhaps the prairies were not meant for Europeans at that time, but what a place they were for animals. In their original state, the prairies of North America acted as an expansive and fertile stage for legions of creatures that fed, bred, and died there. No one now knows the precise numbers, but about 100 million pronghorn antelopes roamed the western and southern regions of the plains. Not less than 60 million buffalo rolled across the prairies in the greatest grass-eating corporation of creatures ever assembled. To these millions were added countless jack rabbits and coyotes, wolves and ravens, quail and grouse, ducks and geese. The biggest population of all—about one billion prairie dogs—was collected in sprawling underground cities. The largest of these cities covered 25,000 square miles and housed hundreds of millions of creatures.

The size of the prairies may have repelled the early Europeans, but the 500 million acres made a variety of environments for the innumerable creatures living there. A man's first view of the prairies showed rolling, silken grass disappearing out of sight. But this was only a tiny part of the entire prairie world. In other areas the prairies might be parched gray-brown and spotted with struggling cactus trying to get started under the Nebraska sun. Country club lawns undulated for hundreds of miles. Jungles of hazel and dogwood, wild plums and hickories, bur oaks and chokecherries spread across some sections. There were also canyons and badlands and near-deserts. Although rainfall might be irregular, sparse, and unpredictable, the prairies also hosted ponds and marshes. These attracted such great numbers of waterfowl that the blue skies were decorated with ducks and geese and swans, which used these patches of waterland as way stations on their migrations.

Once, standing on a hill overlooking Denver, Colorado, I felt the full power of this prairie world. The immense sweep of land, bathed in the reddish glow of a late-afternoon sun, was only a tiny part of an even greater slope of land that began in the Rockies, passed through Denver, and fell imperceptibly away toward the Mississippi.

That the prairies were once a world overflowing with creatures is true enough, but their success depended on the triumph of an earlier form of life. The earth hosted grass before it accepted trees. Grass climbs mountains into snow country, and descends into the salt sea. Grass can withstand years of drought, fire, the trampling of animals, the work of insects, the digging of gophers, the gusts of gale, tornado, and snow. Grass found in this great prairie country had a chance to establish itself after the prairies were formed by the 60-million-year movement east of billions of tons of debris from the slowly eroding Rockies.

Thus, when the buffalo made the plains their own, they were led on endlessly by the lure of these abundant grasses. In the eastern prairies, an acre of tall grass could keep a buffalo alive for sixty days, giving it three thousand pounds of green food. On the shorter grasses of the high western plains, the buffalo could graze out an acre of drought-stricken land in two or three days. The interchange of grass-to-buffalo-to-dung-returned-to-the-earth was a truly colossal transfer of energy. The success of the grasses sent some of them as far north as the middle of Alberta, more than two thousand miles away from the limit of their southern growth near Mexico City, and gave the prairie creatures a generous range in which to roam. The grasses locked together to form dense communities. This stopped other intruders from penetrating their territories. A single grass plant could grow five or six hundred miles of roots in a year.

While the buffalo trudged on in their millions, grazing, mating, fighting, and dying, the grass communities became nations, then empires. In the dry western prairies, which are sheltered by the rain-catching Rockies, the blue grama grass and the buffalo were staples of the buffalo herds. Millions of grazers clipped these grasses to the height of a mowed lawn. At the eastern edge of the prairies, where Gulf of Mexico rains drove north, cordgrass grew so high and dense that a man on horseback could neither see nor navigate through its jungle thickness.

Between these extremes the great communities of grass fought for dominance. Uneasily sharing ground, the big bluestems grew with the side-oats grama; the tall dropseed flourished with the purple lovegrass; the porcupine grass and the silver-leaf psoralea moved together in waves among the crushing hoofs of the grazing creatures.

All the animals were bound in a firm alliance with the grass. The buffalo remained unyielding in their incredible numbers, but the grass was always flexible in its ability to cope with the many seasonal variations. During hasty thaws, grass roots hung on in subterranean safety, while dead buffalo clogged rivers—the victims of melting ice and tumultuous floods. During freezes and howling blizzards, the antelopes starved, but the grass survived, slumbering under the protective blanket of snow. During drought, billions of miles of live grass roots stayed inert, while buffalo, lacking food and water, left their carcasses to rot on the prairie lands.

Despite the scope of such periodic disasters, the prairies were a spectacle of vibrant survival and industrious multiplication. The buffalo and the grass set the scene for this sustained drama of the plains. The buffalo never moved predictably. In good seasons, they traveled only far enough to get fresh food, but when they were hungry or thirsty, they devastated the land, clipping the grasses to dust and then digging out the roots. Scores of thousands of them came to tiny water-holes, crushing sedges, arrowheads, and cattails, smashing down duckweeds and water ferns, and destroying the nests of blackbirds, waterfowl, and sparrows. One thousand different species of life could be obliterated in the one hour it took the thirsty animals to drink. An unsullied prairie pool might take years to recover from such a visit.

Moving in herds that stretched from horizon to horizon, the buffalo overwhelmed the prairie world as they traveled. But a host of other creatures followed them. The gray wolves, sagacious, cooperative, and dauntingly strong in their packs, were expert buffalo hunters. They cornered old bulls or young animals and hamstrung them, or wore them down in long runs. The wolves were always the most visible hunters, but the most ingenious hunter, and perhaps the most numerous, was the wily coyote. It was everywhere, the gray ghost of the prairies. It could hunt 150 miles in a day or a night. It was nearly omnivorous; a berry-eater and a worm-hunter, a killer of nestling birds and grasshoppers. It ate snakes and sucked goose eggs. It did not matter whether the prairie was parched grass or flooded with ponds and marshes. The coyote could find a living digging out voles, or swimming underwater to seize paddling ducks. Coyotes worked in pairs to kill calves or sick buffalo females, and scavenged the uneaten carcasses of wolf kills.

The buffalo expanded to exploit the territory grass had created, and as they fed, so were they fed upon. Billions of flies traveled with them. The flies were so numerous that early prairie voyagers could not get food from plate to mouth without clusters of flies reaching it first. Many of the flies were bloodsuckers. Many sought buffalo dung in which to lay their eggs. Dung was the prize of the prairies, and millions of tons of the stuff poured out each year. Maggots teemed in it. Scarab beetles struggled to get to it first so they could bury it with their eggs.

The bountiful grass sprouted in obedient response to the seasons. It fed the buffalo, the antelope, and all the others, but it gave special power to the grasshoppers. For most of the time, the hundred-odd species of grasshopper were the humble suppliers of food for the mice and other rodents, and for the quail and grouse which populated the plains. The grasshopper lived in a competitive struggle with swarming beetles and parasitic wasps, moths and many flies. But, periodically, the grass-

81. Sky and earth make a powerful image of the American prairie, where the scope of the land gives plant and animal expansive opportunities for life.

hoppers swarmed. The sound of their wings was like the roar of a prairie fire hissing and thundering across a front many hundreds of miles wide.

They settled in the path of the buffalo and stampeded them. They settled among the charging animals, and flocked ahead of them, so that there was no place for the buffalo to find relief from the falling insects. The swarming of the grasshoppers was so great that they could eat out the entire range of grasses, leaving the earth stripped bare. The buffalo's brain was confused by the scope of the devastation, and it stumbled around helplessly. Starvation decimated its herds.

The great success of the grass inspired fecundity among all the prairie creatures. Beyond the grasshopper lay the ubiquitous meadow mouse, a creature so prolific that a female mouse bred only thirty days after its own birth. Mouse populations rose and fell rapidly —from ten mice per acre to more than ten thousand. The white-footed mouse, just as productive, was another primary resource. It could outbreed all its enemies and provided a continual supply of food for fox and coyote, weasel and badger, hawk and owl.

Winds rushed across the primeval prairies unimpeded. They swept away mounds of harvester ants. They dropped into creeks and river beds and sucked up sand like giant vacuum cleaners. They caught horned larks and tossed them away like scraps of paper. Magpies came over the crests of ridges in the western prairies and were driven backward in the grip of the winds.

The wind reported the nature of the next day's weather to all the creatures, and the plains carried a continual code signaling seasons, storms, calm periods, times to migrate, and times to breed. When the first purple townsendia appeared in the western prairies, its daisylike flower, crouching close to the dried grasses of the previous year, announced that it was spring. In the central prairies, the dogtooth violets heralded spring, while pasqueflowers proclaimed the new season in the east.

Each season also had its sounds. The yapping chorus of coyotes, the booming of prairie chickens, the fluted, melodic howling of wolves, the rumbling of buffalo hoofs, the chatter and whistle of massed birds flying overhead at night, the honking and squawking of millions of waterfowl on the move—all these spoke of spring and the fruition of an expansive and abundant land.

Spring was given color by a flood of flowers. The appearance of plants was on such a scale that species mixed with species, each group growing in separate layers, one upon the other. Bluestems rose around the spring flowers to create expansive green carpets. Each day the balance of flowers, shrubs, and grasses changed. Beneath them all waited the autumnal plants; the gentians, asters, goldenrods, and sunflowers.

The ceaseless industry of the plants was matched by the work of the black-tailed prairie dogs. They dug continually into the top dozen feet of the plains' surface, and their burrowing encouraged the growth of forbs, which fed the pronghorns. Prairie dog burrows gave shelter to owls and rattlesnakes, which sometimes ate the dogs.

The adaptable and numerous prairie pronghorn antelope, which was not a real antelope at all but an animal halfway between goat and antelope, thrived here. It had the eyesight of a small telescope and the endurance to outrun any gray wolf, in a race that could go on for hours over hundreds of miles. It had the speed to outsprint a cat at fifty miles an hour, and a cavernous breathing system which allowed it to keep going indefinitely. The pronghorn could live in 110-degree heat, or at 40 degrees below zero. It needed almost no water, so it was free to prowl the dry western slopes of the prairies. It did not compete with the buffalo for grass but ate forbs that sprang up around the range grasses: fringed sagebrush, snakeweed, bindweed, yarrow, locoweed, larkspurs, and even the prickly pear cactus. The pronghorn was the triumphant exemplar of all those animals that occupied the great, original grass pioneers of the prairies.

But, alas, most of these creatures are gone. As late as the 1870's, prairie chickens still teemed in their millions, and a hunter could knock down a dozen of these grouse with one shot. The destruction of most prairie creatures was completed in the latter part of the nineteenth century. The surface of the prairies was scattered with the bodies of ravens, the skulls of coyotes, the piled bones of buffalo, the horns of antelope. Poison, bullets, gas, traps, plows, and bounties ended the lives of the prairie creatures and changed forever the world in which they had triumphed.

Today, there are only echoes. I caught one of these reminders of the past in Nebraska a few years ago when I was alone on the plains. I saw a solitary, high-flying bird approaching from the east, heading toward the sinking sun. I thought at first it was a goose, or a duck, but its rapid wing beats, interspersed with short glides, marked it as a grouse, a prairie chicken. The sun, setting behind me, touched the bird's plumage so that it looked like a small, flying jewel. As I watched, the prairie chicken passed overhead, diminishing to a lonely speck set against the crimson curtains of the west.

82. Exemplary exploiters of the vast western prairie grasses, the American bison once moved across the land in great migrant herds. Two bulls rest in an autumn meadow in South Dakota, bodies stored with fat for the tough winter ahead.
84. The sixty-mile-an-hour sprint of the American pronghorn, an antelope in name only, escapes all but the most tenacious wolf packs. Unlike the bison, the pronghorn needs little water and may venture into deserts and dry mountains.
86. The shy and cunning coyote is the nocturnal scourge of small rodents and rabbits. Its cries on moonlit nights lends a musical counterpoint to the western wilderness.
88. The dawn display of the male greater prairie chicken on the plains of Missouri tells all others of his kind that he has staked out territory and will receive females there. His hollow booming cries, echoing from the distended neck, carry his message far and wide across the plains.
90. Snow geese turn across a prairie marsh, a feeding way station on the route of their migration to northern breeding grounds. Millions of winged migrants share their journey and their purpose.

6 Plains of Illusion

Photographs by:
Hubertus Kanus: Rapho Guillumette Pictures, 95
Edward R. Degginger, 96, 108
Bob Harrington, 98, 112
George Holton, 100, 106
Bob Harrington, 101, 102
John Dominis: Time/Life Picture Agency, 104
Hans Dossenbach, 109
Clem Haagner: Bruce Coleman, Inc., 110

One morning in mid-March I stood on a low hill two hundred miles south of the Equator and about seven thousand feet above the sea. The rain had stopped and the clear air was almost crisp. Before me stretched the great grass plains of East Africa, and scattered across the immense green pasture were Africa's most familiar animals. Vultures circled a family of zebras, and an eagle floated away to the west, its feathers hissing in the breeze. Lions lay supine under angular acacia trees. A solitary, leggy cheetah sat upright on a tall termite mound. Clustered gazelles grazed quietly near the zebras. Behind them, bearded wildebeest lowered their heads and tramped off in single file to find new grass. An unseen hyena chattered from one of the rock islands which dotted the plains.

I focused my telescope, compressing a long, narrow mile of this stunning highland country into quite a different picture. Watching closely, I saw larks, plovers, finches, weavers, mongooses, warthogs, bateared foxes, jackals, ants, termites, flies, and mauve grasshoppers. A tiny cat slid silently from one rock to another. There was movement in the undergrowth where rats, hares, and weasels hid. Storks dropped vertically, using their wings as parachutes. Everywhere there was a quickening of small lives, a suggestion that they, not the larger animals, are really the main influences of this world. And I was soon given proof that this is true.

A lion jumped from the long grass at a leaping gazelle, and missed it by six feet. He fell back, panting. Next, he attacked a zebra, this time getting his broad paws clamped around the animal's neck. He was trying for a suffocating grip on its nose, but the zebra carried the four hundred pounds of lion easily, and when the lion finally slid to the ground, the zebra stamped on his stomach. The lion growled and chewed at the zebra's ears. It was impressive, but not effective. The bleeding zebra wrenched its head free and ran off.

The power and influence of the lion is not what it seems. When the seasonal rains begin, flushing the plains from their months-long drought, the lion catches very few of the great armies of zebra, wildebeest, and gazelle which advance to the plains from their drought quarters in the woodlands. In fact, all the great hunters—cheetahs, wild dogs, hyenas, leopards, and lions—kill fewer than five per cent of the moving animals, and their effect on populations is minimal.

The seasons of these plains consist only of wet and dry. The drought sometimes persists for more than a year, and the wet often brings torrential rains which swamp the light, volcanic-ash soil. The plains creatures endure such extremes of dry and wet, and the weak become part of the paradox of the hungry hunters; they survive in their hundreds of thousands amid legions of the very strong.

As I explored the grasslands with my telescope on that March morning, I realized that the obvious externals were but a mask for the interior life of the plains. An eagle swooped down into some shrubbery near a water hole and scooped up a mongoose. But the writhing creature, gripped by those clenched talons, climbed the eagle's leg and bit through the feathers of its stomach. The eagle dropped the mongoose. Later, the eagle killed a sick fawn, but the creature was too heavy to transport. The eagle fed off the carcass, then dismembered it, flying away with a limb. Immediately, two jackals arrived, but they were driven off by a hungry hyena. A vulture waited until the hyena was finished, then dropped down to eat what little was left of the fawn. When the eagle returned to carry off another piece of its kill, there was nothing to transport.

A hunting dog paused on a ridge. A gazelle fawn lay hidden in the grass at the dog's feet, but the dog did not detect it because the fawn was utterly still and gave off no scent. Its mother had licked it clean at birth, and its own scent glands were not yet functioning.

Comradely elephants appeared at the woodland fringes of the plains, majestic in their passage. Their size, their longevity, and their group compact seem to have eliminated many of the complications of survival. But as I watched, they moved into a grove of trees. Branches were ripped off, and their leaves swished through the air like great tennis rackets. If the branches could not be torn off, the entire tree was buffeted as if by a six-ton bulldozer until it was uprooted and smashed to earth. The uprooted trees lay ready to be eaten, but the elephants took only tiny clumps of choice green leaves. Later, I knew, fire would come to destroy the wrecked trees. Grass would sprout. Busy antelope teeth would nip all the hopeful seedling trees. The elephants were destroying their own browse place before my eyes.

Behind the facade of life on the plains lurk its tiny, deadly moderators. Three hundred thousand antelope entering the plains are susceptible to a lethal virus, and when it enters their bodies, more than half of them will die within a year. Elephants succumb to parasites which eat deep into their bodies. Another virus causes hunting dogs to hemorrhage from the lungs. Not even the thick-hided rhinoceros is immune from harassment. It is attended by thousands of bloodsucking flies, which drink its blood and are transported on its body until they are ready for their next meal. They breed in its dung, and their progeny are always waiting at the rhinoceros lavatory for its arrival. The oxpeckers are some help; they probe every inch of its gross body, entering its ears and forcing their way so far up its nostrils that it snorts to dislodge them. Young lions bait the rhino. Giggling hyenas dart and circle the big animal until it gasps for breath in its efforts to spear them with its horn.

Every strength is tested by a counter-strength. Leopards kill each other in territorial fights along watercourses. Lions eat their own cubs. The tsetse, a fly supremely well fitted to obtain large blood meals from almost all animals at the fringes of the plains, is itself hunted by countless creatures the moment it has finished sucking. Birds chase its bloated body. Robber flies wait in the shade for the tsetse to appear. There is no refuge; even the soil teems with insects and small mammals eager to eat the tsetse before its blood meal can be digested.

The longer I looked out over the plains, the more interlocked the scheme of life became. A small hawk knocked down a lark. A larger hawk commandeered the body. An eagle dropped to get its share, but a baboon drove them all off. A termite came out of its mound to hunt for scraps of vegetation and was plucked up by a bird, which was caught by a gray cat. The cat was later killed by a leopard springing from the shelter of trees edging a stream. A column of ants marched through the grasslands, catching beetles, flies, larvae, termites, even young mice. The onrushing advance of the ants attracted several different species of birds that hunt insects disturbed by ants. Thus, one species caught the cockroaches fleeing the ants. Another sought beetles, while a third species darted after the moths and butterflies which rose before the advance of the ants. Nearby, a small hawk waited its chance to catch one of the birds.

Supreme specialization governed the lives of these creatures, but I understood very little of the intricate details. The flappet lark rose abruptly, brought its wings together to rap the knobs on them in staccato, machine-gun-like bursts of sound. A large beetle also rose, and for some incomprehensible reason, made the same sound.

The long-legged, imperious-looking secretary bird, which has the legs of a small ostrich, the beak of a vulture, and the wings of a stork, began a series of agile, backward somersaults, apparently in celebration of its mating urge.

When a million small finches gather at the edge of the grasslands to breed, they quickly create a miniature metropolis of nests on the twigs of shrubs and small trees. To this place come snakes specialized in swallowing the finches' eggs and nestlings. The snakes swallow the eggs whole. When an egg reaches its gullet, the reptile contracts its body to crack the shell, then lifts its head so the yolk can slide easily past a valve into its stomach. Afterward, the snake drops its head and expels the broken shell from its gullet. Gray cats accompany the snakes to the finches' breeding ground. They capture the trapped birds in their nests at night. Storks work their way through the bird city too, dislodging eggs and nestlings with their long, clumsy beaks.

The termites appear to be safe in their earth cities, but they are attacked by warrior ants, and bitter battles sometimes wreck the intricate social organization of the termite colony.

Contradictions run through the plains' world. Superficially, the plains grass looks like any other grass, but it is adapted to extremes of wet and dry. During times of drought, it withers and practically disappears. What is left may be swept away in quick fires, or cropped to the ground by animals. Yet the grass waits, quiescent, its ability to remain intact as a living entity perfectly preserved because it grows from the base of its leaves, not from its leaf tips as do most other plants. Hours after rain has fallen, the grass begins to sprout again.

Each plant on the plains is adjusted to cope with both the animals and the elements. Red oat grass twists its bristled seed into the ground. The bristles react to rain or fire by twisting the seed further into the ground for protection. Saw-toothed grass pushes a seedling stalk high off the ground, but keeps its leaves lying flat. Antelope or wildebeest or zebra may graze off the stalk, but the bulk of the plant is preserved on the ground. The grass will thrust up another stalk and grow subterranean runners, which will pop to the surface and thrust up their own seeding stems hours after the animals have passed on.

There is intense competition between the plants for space. Annuals, which grow from seed each year, contest the perennials, which grow from the previous year's roots. Even the trees are in competition. Each species uses its own subtle tricks to survive. Some come into leaf before the rains begin in order to have a longer growing season. Others protect their foliage with thorns. Poisonous ants are hidden in the fruit of some trees, which discourage fruit-eaters. Other trees have bitter-tasting bark.

The sturdy grasses and trees of the great plains benefit from the work of omnipresent dung beetles, which fight each other for possession of animal droppings. The beetle puts its egg inside a pellet of dung and carefully pushes the built-up ball of excrement to a suitable burial place. Beetles bury thousands of tons of dung each year, and in the process help to fertilize the grasslands.

When, after a month of watching, I put down my telescope, the scene before me had not changed much. Lions were asleep; wildebeest were still walking in single file. Giraffes had moved among a grove of trees and were delicately cropping the high branches. The elephants had gone, and a leopard had driven vultures out of the tree where its next meal was stored. There was no hint of the hidden reality that lay behind this serene facade; no suggestion that what I saw was irrelevant to what was actually happening. The external view of a relatively few powerful animals dominating multitudes of weaker creatures as they wandered across the peaceful terrain did not resemble the internal orders which truly govern life on these tropical grasslands.

95. The vast grass plains of East Africa support a great variety of animals, large and small—gazelles, impala, wildebeest, zebra, giraffes, elephants, lions, jackals, snakes, hyenas, vultures —a fascinating and dynamic mixture of hunter and hunted. Scattered, flat-topped acacias give shelter and browse to many creatures.

96. Dusk falls around a wary leopard beginning its solitary night hunt. Smaller than the lion, the leopard is a strong, lithe, efficient hunter, which stalks a favored prey, the impala, either through tall grasses or from ambush in a branch over a water hole.

98. Nearly four feet tall, the gawky secretary bird is well adapted to the African grasslands. A peculiar species, its ancestry is not yet known. With its long storklike legs, it can run faster than man. In a chase, it would rather run than fly. It hunts small mammals and is also adept at killing snakes.

100. The zebra's vivid black and white markings are conspicuous by day on the plains. But when the animal is in motion, the stripes confuse the hunter's eyes. The zebra's alertness, nervousness, keen hearing and sight, and strength mean it is always a formidable adversary, even for a lion.

101. The gerenuk is a gazelle that is evolving along the lines of a giraffe. It can live in drought lands where food is scarce, and it need not drink for months. It can rear up on high, pointed hoofs that give it balance to get at leaves beyond the reach of most other browsing animals.

102. White-bearded gnu, also known as wildebeest, are often found with zebra, a few of which are visible in the background. The two animals will migrate to reach new grass and escape drought. The wildebeest, despite its cattlelike body, is an antelope.

104. Two female impala leap in tandem. Abrupt bursts of speed are their only defense against predators. These graceful animals can make leaps eight feet high and twenty-five feet long. They spurt to nearly fifty miles an hour to dodge leopard, lion, or wild dog.

106. An elephant, its hide colored red by a recent dust bath, threatens to charge. A full-grown elephant cannot be killed directly by any enemy except man. But drought is its worst enemy. It needs as much as 600 pounds of grass and browse in a day. It destroys trees to reach the tender, upper leaves.

108. The short-sighted, bad-tempered rhinoceros charges at danger it cannot even see clearly. But when its charge is accurate, a full-grown lion may be tossed high in the air and killed instantly. The rhinoceros is a solitary animal, moving along its feeding paths from one eating place to another.
109. Three scavengers—spotted hyenas, vultures, and a jackal—crowd around the half-eaten carcass of a wildebeest on the African plains. The collective persistence of these scavengers may be so intense that lions at a kill may spend more time driving them away than eating.

110. A family of giraffe at a water hole carefully scan their world for danger before drinking. To reach the water, a giraffe must spread its legs wide apart. In this clumsy position it is vulnerable to attack, usually by young lions. But out on the plains, its height gives it a long-distance view of all dangers.
112. A pride of lions sleep off a big meal. They eat irregularly and must gorge when big kills are made. But prides remain in fixed territories, and they go hungry if their prey moves away. The lions defend the territory but the lionesses do most of the hunting.

7 The Thrifty Desert

Photographs by:
Ed Cooper, 117
Anthony Bannister: Natural History Photographic Agency, 118
Karl H. Switak, 120
Clem Haagner: Bruce Coleman, Inc., 121
Anthony Bannister: Natural History Photographic Agency, 122

In the Sahara, plains of gravel march for a hundred miles and end in walls of sand five hundred feet high. In the Arabian desert, dunes flow like rippled waves from the Mediterranean to the Euphrates. In southern Algeria and northern Chad, bleak stony ranges lie under blazing suns, and snow-capped peaks shimmer in a haze of heat. When the harmattan blows in the Sahara, it smothers entire countries in a blanket of sand. I have sailed through Sahara sand a score of miles off Africa's Atlantic coast.

In the minds of many human beings, deserts are synonymous with death. But the seeming desolation of the world's dry places is an illusion. The apparently inanimate desert wears a mask to hide its secrets; life is preserved there in ways still beyond our ability to understand. Desert creatures reveal the ingenuity of life forms to beat the extremes of heat and drought. A solitary blade of grass rises from a sea of Sahara sand; a single cactus sprouts in an empty Sonoran landscape; the flat and desolate Australian desert hosts hidden marsupials, and brilliant parrots fly to distant water holes in this sere and lonely land.

As life evolved and emerged from the wet and the humid and the warm, it most likely spread slowly into drier regions. Water was always the key to survival, even in deserts. No desert anywhere is completely dry. Rain falls spasmodically, even though fifty-year droughts may afflict parts of the great Australian desert. The rare rains must be preserved, despite evaporation which would be measured in feet per year if heavy rains did fall. Thus, when rain comes it can be stored only if it can be absorbed quickly into the parched earth, before evaporation begins. When the water gets down a foot or so, it can be retained to serve life for a few months. But if it can sink twenty or thirty feet, it may provide water for many years, perhaps for centuries.

It is on such tenuous connections with water that life has managed to conquer every one of the earth's deserts. Bacteria and fungi thrive in regions where rain has not fallen for years. The Sahara cauliflower forms greenish, mosslike heads which absorb sand and become as hard as stone. I have tried to break one open with a knife, and failed. Other plants protect themselves against grazing animals with thorns, with bitter-flavored foliage, or with unpleasant smells. Some annual plants survive by casting down droughtproof seeds, which can wait for rain as long as one hundred years. They spring to life only when the correct amount of moisture has fallen. The seeds appear sensitive to the quantity of the fallen rain. Perhaps they have some growth inhibitor which is washed from their exteriors by the rain. If they sprout, they grow rapidly into leaves and flowers.

Stored moisture keeps these plants going, and the ingenuity of their water-extraction techniques is directly connected to the more dramatic examples of animal survival. A sudden downpour in the Mojave Desert creates pools of 100-degree water which almost immediately teem with shrimp hatched from eggs laid perhaps a quarter of a century before. Like the annual plants, the shrimp quickly reach maturity.

Spiral-horned, donkey-sized antelope wander the central regions of the great Sahara dunes. The addax drinks rarely and matches the camel's capacity to keep its blood stable while drawing water from other parts of its body. Two gazelles, the dorcas and the rhim, move quickly over long distances to span scanty supplies of food in practically waterless territories.

Each life form has developed such precise adjustments to the immense contrast between the dry and the wet that the desert is a place of persistent miracles, at least in the eyes of the human observer. I remember with delight the flush of new flowers after rain in the Australian desert. It is one of the world's great spectacles; hundreds of miles of blossoms springing from earth which has not been damp for a score of years. I have been equally entranced by the seas of yellow dandelions and mauve sand verbenas which wash across parts of the American desert after rain.

Each desert creature works at minute variations on the common theme of conserving moisture. In the huge sand dunes where all seems lifeless, birds, beetles, lizards, mice, and rats thrive. Desert cats, jerboas, and fennec foxes hunt at night so they will not unnecessarily lose a drop of their body moisture. They have also developed almost perfect resistance to the sting of the scorpion, a creature which can thrive in the hottest deserts. Its large claws can dig rapidly. Its sting is uniformly deadly against the desert insects it hunts. It is nocturnal, but it can stand great heat.

The antelope jack rabbit dissipates the heat of its blood in thin, eight-inch-long ears to conserve its body moisture. Snakes and lizards, nonsweating and slow-breathing, reduce their vital functions to such a low ebb that their bodies conserve almost all moisture. They expel waste products in solid, dry masses, every particle of moisture extracted before evacuation. The kangaroo rat conserves moisture by rarely urinating, and its nasal passages constantly condense very small amounts of water from the hot, incoming air.

The collared peccary, a small American desert pig, solves the moisture problem by uprooting cactus plants, gobbling down the spined leaves, and chomping the roots. The roadrunner, a long-legged bird, eats snakes, and so gets the moisture they have so carefully conserved. The gila monster stores moisture in the fat of its tail.

These animals live among plants that have their own ways to conserve water. The annuals compress the growing season and race to brief climaxes. The speed of their growth helps to counter the eager, searching lips of the many grazing animals. The perennials are much more deliberate. They cut their foliage to absolute minimums, toughen their leaves to reduce evaporation, spread their roots widely, and develop thick, porous stalks to store water. When the rains come, they drink and hoard as much water as possible to prepare themselves for the next drought.

The adjustment to drought is so precise that rain may bring disaster. The sun may shine for a thousand days in a desert drought, each hot, dry day the same as the last. But suddenly clouds gather, the sun disappears, and the rain pours down. In minutes, the sand is gullied by miniature rivers. Plants with fragile roots in the sand are swept away. Beetles concealed inside the sand are drowned or smothered. Lizards sheltering in the shade of rocks are caught up in the torrent and sent spinning away.

Then, the rain stops and the sky clears. The sound of racing waters diminishes. In as little as twenty seconds, the sand is dry. This brings another hazard to those creatures which fled the water, or were swept away by it from their safe refuges. Now, so well displaced, their near-escape from drowning may bring them quick deaths in the heat of an alien place where they can find no shelter.

When the offshore Humboldt Current occasionally warps away from the coast of Chile, rains devastate the coastal desert and the dry foothill country, causing damage to the fragile life systems there that may take scores of years to repair. I have been in the American Southwest when torrents of water poured down dry creek beds after foothill rains and swept almost every living thing away with it, including livestock, people, and houses. Heavy desert rains in Australia catch and drown wallabies and other small marsupials, and form large inland lakes, which may endure for years. I have watched the creatures that come to these lakes to thrive along their shores, and know they must disperse or die when the water declines and finally disappears.

In the areas of truly devastating rains, there is an ultimate desert paradox: the force of the life-giving rains makes practically all life impossible. The Sahara's *hamadas*—plateaulike lands surfaced with stones, rocks, and shale—seem to me like the beginning of creation. There, the rare rains do not sink in at all. The water cannot percolate into the ground to form reservoirs which would help to sustain life until the next rain. Instead, it runs off, racing toward the lowest ground. Any plant or animal attempting to survive on this barren landscape must eventually confront the destructive results of an inch of rain falling in a *hamada*. The water is funneled into an ancient watercourse, or valley, and there may become a score of feet deep and hundreds of yards wide. It rushes along to the sound of grinding rocks and shale. The mass of debris may consist of a million tons of silt and rock and gravel. It spreads wide to form stark, lifeless plains called *regs*.

By comparison, sand is beneficial to life. It is plastic, granulated, cool in its interior, and it absorbs water. It is a vital reason why life is possible in deserts. In some areas, sand's plastic quality resembles that of the sea itself, and creatures live in it much as they do in the sea. The largest dunes in the Sahara would cover all of France, and I have watched in fascination as the prevailing winds ripple them into ever-changing curves, steep walls, and deep valleys. They are like dunes everywhere; shallow-sloped to windward, steep-sloped on the lee side. The dunes have a life of their own, and can be dated by their appearance. I have climbed a young dune, its grains of sand freshly wind-borne and gleaming white in the sun. It was still resisting the forces trying to break it down. But as it aged, I knew, the iron compounds inside the sand grains would oxidize, and the dune would eventually take on the golden glow of maturity.

Sand is also a tablet on which is recorded the normally unseen life of the desert. The snake sidewinds along, leaving a precisely delineated diagonal trail as its spoor. Beetles leave complex, multifooted trails, which look like embroidery work on a giant's cloth. The antelope trot through the impeding sand, safe there from lion, cheetah, leopard, wild dog, which cannot hunt on such loose and treacherous footing.

It is to the sand that men have been led in their attempts to conquer the desert. When I fly over the Arabian Desert, I can see how painstakingly it has been invaded by human beings, but they are reduced to the significance of bugs by the great expanse of sand around them.

They have dug many catchment drains in the sand, but the drains will remain dry for years until rains flush water into cisterns, reservoirs, and tiny fields. Their works, geometric and precise in the vast landscape of curving lines, look like the marks of a fossil people. The men have dug *foggaras*, intricate underground water drains to reach the stored water and carry it for miles to stimulate the growth of oases, networks of fields, groves of palm trees. Thus, men respond to the desert in exactly the same way as the animals and plants do. The prime function of all life in the desert is to conserve, to reach the water, to direct and store the precious moisture.

Like the polar regions, deserts give life an outside chance for existence. Creatures which inhabit these dry places show us there is practically no limit to the capacity to survive, as long as there is oxygen, some moisture, and a little food. Each desert displays a separate personality, and in this respect deserts closely resemble the seas. But only one desert, the Namib in southwest Africa, depends on the sea for its own unique life.

The Namib is the most hostile of all deserts. Its sand dunes roll for hundreds of miles without interruption or change. There is no water, no rain, no vegetation. Yet this desert swarms—in a desert sense—with life. A large number of species of tenebrid beetles have made the Namib their home. They use their flattened bodies and large, triangular heads to "swim" through their sandy world, thus avoiding the killing heat of midday.

Although the beetles live scores of miles inland, they are connected by an unbreakable chain to the life-giving, and life-creating, sea. The cold Benguela Current flows north through warm south-Atlantic waters and stimulates a constant suffusion of planktonic life along its fringes. These tiny creatures and plants are driven ashore, dried, picked up by the prevailing wind, and carried inland over the dunes. The portable food is well spread, and the beetles, hunting at night to absorb dew into their bodies and avoid the heat, feed on the tumbling plankton.

I remember the Namib whenever I travel across deserts. All deserts diminish man's dreams. They overwhelm his Roman cities, his Australian ranches, his American mines, and lap at the edges of his pyramids. They demonstrate a paramount truth: The ingenuity of animals surviving in this dry world has not been equalled by man, and never will be. As the works of man are impartially wrecked and buried by the sand, we must remember that beetles living in the Namib eat plankton from the sea.

117. Deserts may be sandy, arid, and hot. They may also be cold, as in mountains, or wet, as in the great tundra lands around the Arctic. They can be stony, as in parts of the Sahara, or extremely fertile, as in parts of Australia. The floor of Death Valley, California, is seen here in its hot, parched desolation, well below sea level and shut off by mountains from moisture-laden Pacific winds. But animals are found in all deserts, including the blazing hot Mohave in the American Southwest, the rolling sand dunes of the Namib in South-West Africa, and the Great Sandy Desert of Australia. Reptiles, mammals, birds, and insects have all made ingenious adjustments to survive in the harshest desert conditions.

118. An adder obliquely loops its body to move across the loose sand of the Namib Desert in South-West Africa, an efficient means of travel also used by the sidewinder of the American Southwest.

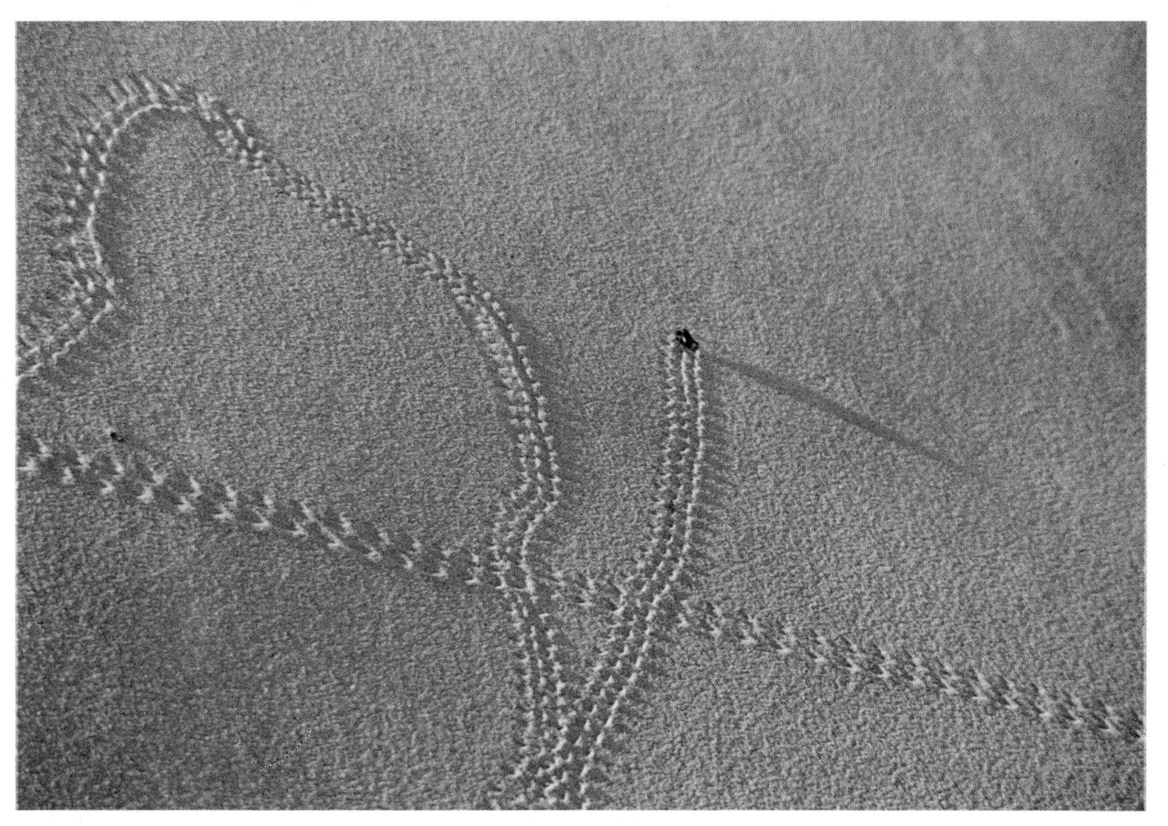

120. A ground beetle leaves an elaborate track in the sand as it moves over the desert dunes near Palm Springs, California. The insect life of a desert is rich but scattered. Many eggs and dormant pupae lie waiting for the occasional heavy rain that will stimulate them to new life. Desert insects may be able to withstand great heat, freezing cold, long drought, and heavy rains.
121. Two gemsbok race across the sand dunes of the Kalahari Desert in southern Africa, their magnificent four-foot-long horns a distinctive mark of these antelopes. They have made the desert their home and can travel far to reach the small stores of water that are available.

122. A nocturnal scorpion of the Namib is one of the great hunters of this vast and inhospitable desert. Threatening, poisonous, wide-ranging, it is a scourge to all the insects, particularly beetles, which emerge from the sand at night to feed. It may eat its own weight in insects nightly.

8 The Diverse Woodlands

Photographs by:
Ed Cooper, 127
Erwin Bauer, 128
Hans Reinhard: Tony Stone Associates, 130
Robert Carr, 131
Larry West, 132, 133
Jean-Louis Frund, 134
Eric Hosking, 135
S. D. MacDonald, 136

At the beginning of winter, with thick snowflakes dropping silently through bare branches, the New England woods seemed to have become a giant dormitory. I walked through frozen summer vegetation, deadwood, earth, mud, and all, I knew, were harboring an immense community of animals which, in my anthropomorphic eyes, were sleeping. The tunneled soil was filled with dormant earthworms. Carpenter ants were buried in the wood of maple trees. Other ants, comatose in subterranean burrows, were clotted into dense masses, some gripped by ice crystals. Ladybirds were jammed together—about twenty thousand of them—in rock crevices. Frogs, toads, and snakes were held, quiescent in earth, litter, and mud. Fertilized queen hornets waited in shelter for spring, along with dormant bumblebees. Other bees waited inside goldenrod stems. Spiders used old birds' nests. Bluebottles had protected their bodies from fatal freezing by burying themselves in the porous insulation of rotting wood and going to sleep.

A white-tailed deer stepped into a clearing. Snow had collected on his back, and he shook himself. Under his feet snails slept, chipmunks dozed, and millions of seeds rested in readiness for the spring race to life. The buck saw me and was gone in vaulting flight. I looked up and met the bright eyes of a silent cardinal, its crimson feathers ruffed in outrage at my presence in this refuge world. A cottontail rabbit bounded past me. Later, it would sink in the deep snow, while the snowshoe hare would travel safely on broad feet.

The winter sleepers around me were merely an underpinning of the woodlands, a tiny fragment of the life that would swarm here when the snow melted. I leaned against the gray trunk of a beech tree and thought of other woodlands I had explored: a topknot of oaks growing on a hill in Leicestershire; a glade of pines in a Norwegian fjord; a cluster of sugar maples in Quebec. And somewhere in Siberia there were expanses of spruce I might never see. All these woodlands are triumphant survivors of the weight of ice that once smothered their territories. All are products of millennia of change; of species fighting species; of animals victorious one millennium and extinct the next.

For me, the woodlands are symbols of an eternal condition; life triumphant in a parade of death. Whenever I walk in the woods, it is this continuum of struggle that I see and feel rather than the apparent serenity and changelessness of trees growing quietly in the sun.

That winter in the New England woods suggested the interconnected system of life still going on under my feet. Pursued by weasels, mink, owls, and hawks, mice had been tunneling through the snow to harvest frozen seeds. A partridge, buried in a snug ice chamber during the snow storm, had been smelled and caught by a hungry fox, which had earlier missed a rabbit.

The web of woodland life is vast and precise. If the fox does not kill the rabbit, the rabbit will kill seedling trees. If the cougar, or the wolf, cannot kill the deer, it will over-browse young trees and ring-bark older ones. If a species of insect is missing in the woodlands, the entire forest may be unable to seed, or flower. If certain insects are absent, they may make some trees vulnerable to fatal attacks by a virus, a fungus, or bacteria.

When the woodlands awaken in response to the urgings of the spring sun, this interconnecting web is strained by the arrival of millions of creatures rousing themselves. Spring peepers kick out from resting places and pipe their calls in shrill voices. Box turtles break the crusted soil above them. From wet leaves come centipedes and bedraggled flies; from hollow trees, mourning cloak butterflies, bats, and moths emerge. And the migrant birds pour in—among them sparrows, phoebes, warblers, bluebirds, and grackles—while the rising sap is drawn up from the roots through the veins of woodland trees until it reaches the tip of every branch and pushes billions of buds from their protective sheaths. The flies, bees, and gnats, still sluggish in response to the cold, drink from the blossoms of skunk cabbages, one of the earliest plants to appear. After the skunk cabbages come a rush of countless other plants, such as bugbanes, violets, rue anemones, trilliums, and toothworts.

I have shared these spring moments with the woodlands, and it seems to me that mankind, too, must have developed here since I cannot think of any other place where the common purpose of life is so well integrated and displayed. Dominion over each square inch of territory is contested in the woodlands. Migrant birds appear, but they cannot settle where they please. They must sing and fight until they have won a place for themselves—and perhaps not more than half a dozen nesting pairs can occupy an acre. The young raccoon, now awake and independent of its mother, is not allowed to linger in another raccoon's territory, nor, for that matter, is any squirrel, chipmunk, woodchuck, fox, opossum, or mouse permitted such intrusion. Every part of the woodlands is filled with this push of life, this total engagement with the new season.

When a Cooper's hawk speeds with uncanny accuracy among tangled branches to clutch a goldfinch, its outspread wings sprawl over several thousand tiny creatures working to transform the debris of the woodland floor into soil. Indeed, without their constant activity, the forest would eventually choke to death in the detritus of its own past successes.

It is a delight to walk almost anywhere in the spring woodlands because so much is happening. It does not much matter whether I happen to be in Sweden or Scotland, in Maine or Oregon. Almost certainly the frog choruses will be sounding, turtles will be mating, muskrats and hares will be copulating, thrushes will be gathering mud for the cups of their nests, finches will be courting, and insects will be emerging from their innumerable places of refuge.

It is a time when creatures are so preoccupied with the tension of the mating season that they lose their customary caution, and the woodland predators enjoy good hunting. The handsome towhee mates in the undergrowth, and is later killed by a small hawk. The bumblebee drifts through a sun-dappled clearing, then is struck to the ground by a robber fly. The armored beetle scuttles to safety, but a fly with a cutting tongue bores a hole in the armor and sucks out the beetle's body juices. Crows patrol the woodlands searching for ill-placed nests, unwary mice, or a squirrel's hoard of nuts. The owls sense the opportunities and come out to hunt in daylight. Once, walking in a Canadian wood, I saw a dozen owls in an hour. They looked at me, unmoving and unafraid, and I knew how much of an intruder I was at this time of opportunity.

But all these visible activities are merely surface details in the great expansion of woodland life. Most of the action remains unseen. It is performed underground, in the wooden corridors inside trees, and in their high branches. At night, the invisible pressures increase. The noise of katydids muffles the staccato bark of a fox. The boom of a great horned owl, followed by the abrupt screech of a stricken bird testify eloquently to the dangers of the dark.

During the day, motionless katydids stay silent and unnoticed in leaflike poses against green foliage. Sphinx moths camouflage themselves on the bark of hickories, and even the great green luna moths escape attention when they lie still on algae-stained bark.

The spring activity in the woodlands leads to the production of eggs, or their equivalent, and the miracle of emerging life can be witnessed by anyone patient enough to watch. Mole cricket eggs are laid in underground nests; wasp eggs are delicately placed on drugged victims, in huge paper nests, and in mud containers; polyphemus moth eggs are laid under leaves; ichneumon flies drill through wood and lay their eggs in the bodies of hidden grubs; stick insects pour eggs by the millions from their abdomens. The greatest producers of all—the trees—flower and begin a new generation of seeds that will try to find and penetrate every available growing place. But only one tree seed in a billion survives. That, I suppose, is the real point of the woodland drama.

Death is the inevitable governor of this rush of woodland life. The nighthawk, crouched in camouflaged perfection over its eggs on the woodland floor, must face the vulnerability of its nest, as must the grouse, booming its ventriloquial call on coniferous woodland slopes. And when a wood duck's youngsters leap from their tree-trunk nest, hunters await them on the ground and in the water.

With the eggs all laid, and with summer well established throughout the woodlands, life consolidates its gains. The young are tested to find those suited to survive, and in the heat many creatures eat less and reduce their activities. Some even go to sleep. With reproduction completed, there is no point in needless work. A period of rest is necessary for some creatures so that they may become ready for the next flood of eggs which will be poured into the woodland world.

The whole purpose of the fruitful summer is to enable the stand-fast creatures to survive the coming winter. That is the evolutionary point to resident life in the woodlands. Raccoons must fatten themselves to prepare for winter sleep, as must the other hibernators. The female bear goes into hibernation pregnant, delaying the development of her young so that they will be born at an opportune moment in late winter.

Other creatures lay eggs that will not hatch until the winter is over. Black crickets force their flexible abdomens deep into loose patches of earth and deposit eggs there. Katydids fasten flat oval eggs along the edges of leaves and twigs. Walking stick insects drop eggs among leaves where they are covered later by protective snow. Plant lice bury their eggs in bark. Leaf and tree hoppers hide them in grass. Aphids place their eggs everywhere in the bark of trees, unintentionally providing food for nuthatches and chickadees.

Once the frosts of fall begin, the flight of the migrants goes into full swing, and the exodus is fully as exciting to watch as the spring arrival. Resident beetles, leafhoppers, ants, toads, salamanders, and turtles head for the safety of the soil. Raccoons must choose the proper winter dens or face an uncomfortable sleep. The woodlands are filled with sluggish hornets and wasps wandering aimlessly as they await death. Their species' survival is ensured by their fertilized queens, which will sleep through the winter.

When the leaves have all been blown away by the last of the autumn winds and the ground birds, such as the quail and grouse and pheasants, become visible, the woodland year acquires another meaning for me. Standing among the stripped branches, I become aware that a prime source of life on earth is trees. In their subtle interaction with the sun and the water and the soil, countless trees, with their canopies of leaves, manufacture oxygen, prevent floods, create and enrich soil, help to retain water in the ground, and provide shelter for every form of life living among them. Walking alone among the bare trees, I am reminded that I am an animal, and that I owe my existence to the woodlands of the world.

127. The woodlands of the Temperate Zone are among the most hospitable of all zones of life. Few extremes of desert heat or arctic cold strike them. They are mostly well-watered and sheltered and so give refuge and food to a rich variety of mammals, birds, and insects.

128. The great regulators of the woodlands, the American white-tailed deer, browse fresh growth and so control the density of undergrowth. This, in turn, helps dictate the growth of young trees; the numbers of white-tailed deer must be in balance with this growth, or else both woodland and deer will suffer. Man's rifles now control the deer populations in place of long-gone wolves and cougars.

130. Fallow deer, beautiful small creatures with a rich reddish-brown coat spotted with white, are found from Asia Minor through Iraq to Iran. Their large, widespread antlers start growing—in the males only—late in spring and are fully developed by fall, when the velvety skin covering, as shown here, is scraped off. The antlers themselves fall off early each spring.
131. The common raccoon thrives even in densely occupied urban areas. It has rapidly extended its range from Central America to Canada. It moves slowly but climbs and swims well, is tough and persistent, and its dexterous hands can open man's garbage cans and doors. Raccoons can be tamed when young but may not remain tame.

132. The red squirrel harvests cones in the fall and stores the seeds, sometimes creating piles of cone scales several feet deep. It is a creature perfectly fitted to the woodlands and is found throughout coniferous forests in Canada up into Labrador, and also in the mountains of the eastern United States. Its bulky nests in trees and its chattering cries are charming additions to many mountain pine areas.

133. Versatile land snails, which can dig deeply into the ground to estivate, have eyes mounted on the tips of retractable tentacles. These turn outside-in to give extra protection to the eyes when danger threatens. But if a tentacle is cut off, the snail regenerates a new one with completely restored sight.

134, 135. Excellent woodland hunters are owls; unseen and largely silent, they find their prey—small animals—with exceptional night vision, acute hearing, and heat sensors, which enable them to seize a mouse even when it is buried in thick grass. Their hooked claws and beaks are used in combination to attack. Like hawks and other birds of prey, they can survive long periods without food. The boreal owl (134, left) roams the north woods and hunts as well in blizzards and biting cold as in hot weather. The tiny saw-whet owl, only eight inches long, shares the boreal's habitat in summer but migrates south in the winter. The barn owl (135), a secretive creature that once lurked in hollow trees and caves, now hides and nests in farmers' barns.

136. The blue grouse pauses in its coniferous woodland habitat on a western mountain. A typical plump-breasted grouse, it has short, rounded wings and colors that blend in with its background. About eighteen inches long, it feeds on the needles of conifers.

9 The Stable Rain Forest

Photographs by:
Karl Weidmann: National Audubon Society, 141
Douglas Faulkner, 142
Edward R. Degginger, 144
Karl Weidmann, 146
Edward S. Ross, 147
Karl H. Switak, 148
George Holton, 150

Once I found myself caught in a monsoon rain far from any shelter in a Malayan forest. I remember feeling utter helplessness as the forest roared with the sound of falling water. More than a hundred feet above my head dense rain struck the upper canopy, then solid torrents of water rushed and hissed down the sides of trees as though a thousand large water mains had been turned on above me. The effect was frightening, and the irrational fear of drowning was uppermost in my mind. No collection of plants is better named than this: the rain forest.

I understood nothing then of the marvels of the rain forest. It seemed luxuriant and rich, so I did not suspect that it grew from impoverished earth. The rains were so great that they had long since flushed most of the nutrients out of the forest soil. I did not know that through its millions of years of evolution the creatures and plants of the rain forest had perfected many methods of capturing and storing essential potassium, phosphorus, nitrogen, and calcium to prevent their being flushed out into the ocean.

I did not know that under my feet lay a widespread network of fine tree roots, capable of almost instantaneous absorption of the food brought to them from above. I was yet to learn that the rain forest is a closed cycle of life, making it self-perpetuating, stable, and nearly changeless.

In the rain forest, almost nothing is wasted. Each falling twig is exploited by fungi, bacteria, or termites. The calcium is extracted from leaves and branches, so that none of it reaches the tumbling streams taking the rain water away. The leaves of the forest capture nitrogen, phosphorus, and iron, literally plucking it out of the air.

When the rain had stopped, the enclosing network of lianas became visible high above me, twining through the smothering canopy foliage. The secondary trees and the shrubs grew so thickly that I felt I was walking inside a dimly lit and enormous cathedral. The air had become still. The silence was complete except for the faraway wind rustling high branches. It was like the sound of distant surf.

In this dripping, gloomy, almost silent world is contained the most ancient of all earth's life systems. There, I was very close to ancient geological ages when much of the humid, swampy world was dominated by dinosaurs, giant ferns, and flying reptiles. The rain forest has not changed much in the last 60 million years.

There are no large-scale epidemics in the rain forest, no great catastrophes. Beautifully integrated mechanisms control populations, keep out foreigners, prevent plagues of caterpillars or any other kinds of creatures, stop flushes of fungi diseases, prohibit explosions of parasites.

This stability would seem to favor the survival of a few of the fittest creatures, but precisely the reverse is true. A large rain forest, particularly the one in the Amazon region, may host six hundred species of birds, four or five times as many as a forest in North America. Species of insects run into the hundreds of thousands. Stability does not bring conformity. Although rain forests once extended almost unbroken around the waist of the world, ten degrees north and south of the Equator, they contain no homogeneous groups of plants or animals. The rain forests of South America and Africa share practically no animals, and very few plants of the same species. There are no hummingbirds in Africa or Asia; no orangutans, rhinoceros, or black leopards in South America.

Stability, the rain forest suggests, creates complexity rather than simplicity. Here is a system of life immensely complicated. Almost all niches are filled, yet the life system is fantastically efficient. Billions of consuming and decomposing organisms keep the floor almost clear of debris.

The forest has been developing for so long that practically every order of plant has now evolved into a tree. The bamboos, which are grasses, can grow eighty feet high, sprouting three feet a day in good times. Milkworts have become trees. Violets grow as big as apple trees. The rain forest is a study in elongation. The push of life is upward, and irresistible in its insistence. The result is a repetition of trees with long, slender trunks topped by a grand sprouting of branches and foliage. They triumph over the next group of trees, which cannot spread their crowns so wide and therefore are more closely packed together. The great lianas, some of them as thick as a man's waist, move from tree to tree, binding them so they do not fall even after they have died. Below the second tree level is the extremely dense layer of smaller trees, which stand between twenty and thirty feet high. On the bottom layer thrive herbaceous plants and trees struggling to reach higher levels. Each layer of the rain forest has its own special community of insects, animals, flowers, and other plants, and is the scene of intensive competition for food and space and light. On the forest floor, army ants cover every inch in a ceaseless search for food. They are followed by two hundred species of antbirds—active and eager starling-sized creatures which do little except prey on the insects disturbed by the foraging ants.

If the interior of the rain forest seems quiet and lifeless, a trip through the canopy, or along the edge of the forest, brings another impression. Thirty species of hummingbirds teem among the flowers and fruits produced in abundance, along with two hundred species of tanagers, honeycreepers, toucans, parrots, and cotingas.

There, the colors of the forest grip the eye. Gorgeous yellow parrots place themselves on somber green leaves to create sharp contrasts. The colors of the tanagers are startling; exquisitely subtle shades are mixed with bold, bright hues, so that when these birds are seen against the flowers and fruits of the trees, they produce ever-changing patterns. Butterflies decorate the forest everywhere. They pause on leaves, one ray of sun touching the beautiful golds and reds, the blues and purples of their wings. Beetles, another vast group of rain forest creatures, show other, more muted colors. Nearly all the brightly colored creatures have developed bad odors or foul-tasting bodies to protect themselves from hunting birds.

The rain forest is an evolutionary showcase which hints at the way life was in the deep past. The sloth is one of the world's slowest creatures and it has capitalized on its unlikely talents. It spends its life on a branch, hanging upside down from two- or three-clawed feet, its food—leaves and fruit—within easy reach. It is rarely seen by its enemies because of its slow movements and because it is infested by green algae which tint it to resemble moss.

The hoatzin, a clumsy, chickenlike bird, nests along the shores of rivers meandering through the rain forest and produces nestlings which still have claws on their wings in testimony to their reptile ancestry. At dawn, the deep-throated growling of the howler monkeys begins in the Brazilian rain forest, giving the scene a primeval quality. The cries gradually change in timbre until they become great hollow roars, which, in concert, can be heard a mile away.

American monkeys are a diverse group of creatures which may go through their entire lives without ever touching ground. The marmosets, graceful little animals that flit through the upper branches, trail long tails and utter twittering cries. The saki sits motionless on a slender branch, its extended tail resembling a cluster of dead leaves. The uakari—its body smothered in russet, bushy hair—gives out peals of laughter, blushes red when disturbed, and flees through the treetops.

The American monkeys have made the rain forest their own and sail through its high branches as if they were truly flying. But the apes, which have developed beyond South America, are less certain in high trees. Some of them have invaded flat and open ground, and have evolved group defense against common enemies.

With its great ages of stability, the rain forest has hosted many creatures which have evolved beyond its boundaries. But it also continues to offer a haven to those which seem to have slowed their evolution. The peculiar marmosas reside in the branches of the tallest trees, literally at the limit of their environment. These rat-sized marsupials are so secretive in their nocturnal habits that we know little about them. But they are successful; more than thirty species flourish in the rain forests of South America. They are peculiar because they do not have pouches like other marsupials. Instead, their young cling to their mother's teat unprotected.

Life in the trees causes many kinds of parallel evolution. All the opossums of

South America, except one, have prehensile tails. So do most of the monkeys. The tails are useful, even vital, to help creatures living in territories which may flood regularly and force them to move entirely in the trees. One opossum, however, has elected to face the flood waters on the ground and has become partially aquatic. This water opossum has webbed feet and dense fur, and it can dive for frogs, shellfish, and other aquatic creatures.

The rain forest is an ideal world for reptiles and amphibians. The temperature varies only about four degrees throughout the year, so that the cold-blooded creature, which is at the mercy of the weather, is safe. The rain forest offers countless places to hide. The forty-foot-long anaconda, draped over a greenish branch in the gloom, is almost invisible. The caiman, a relative of the alligator, drifts slowly through water below the anaconda and is watched by its greatest enemy, the jaguar. An emerald tree boa melts into the foliage.

Every tree has its tree snake. Many of them are long and slender so they can reach from one branch to another. Some are poisonous; others are constrictors.

It is ideal country for frogs. Most of them can climb, and many spend their entire lives in the treetops. They breed in pools of water collected in hollows, plaster their eggs to the bark of trees, and produce youngsters from eggs stuck to their backs or concealed under their skins. But they are still vulnerable in high places, and some of them have taken deadly defensive action. They are brilliantly colored—orange stripes set against gleaming black, red, and yellow—which advertises that they are poisonous. The poison is concealed in skin glands and is released on contact. A tiny drop may paralyze a hunter when it enters the bloodstream.

The insects are so stratified that yellow fever mosquitoes that roam the treetops may never come close to the ground unless the trees sheltering them fall.

The world contains hundreds of species of mammals, but there are more than four million species of insects, and the rain forest probably contains at least a million of these, though most of them are as yet unidentified.

The canopy swarms with moths, many of them as large and gaudy as butterflies, and with the fast flight of birds. Real butterflies—the morphoes—are huge and beautifully colored. They, too, fly in the highest parts of the rain forest. Some fifty species show brilliant colors to the tropical sun. The owl butterflies, however, are dull colored, skulking in the dusky shade.

Because the rain forest is so well exploited by its creatures, every possible camouflage and stratagem is needed to survive. The katydid blends almost completely into its landscape, wings showing what appear to be spots of fungus, or sparkling droplets of rain. Some insects have wings that look like leaves, and the camouflage is so perfect that portions of the wing appear to be eaten away. One frog looks like a bird dropping. The African rain forest leopard is black and near to invisible in the shadows. The jaguar of the Americas has grown to resemble dappled light falling through canopy leaves.

Ants and termites, those ancient forms of life, do their work in the great rain forest. The termites consume wood, quickly disposing of waste. The ants cut entire leaf crops off some trees. Workers clip away at each leaf until the tree is bare. Other workers on the ground gather up the fallen leaves and transport them back to the nest where they are chewed to a pulp. Afterward, they are put into a compost heap in which the ants grow their food—and fungus. Ants climb the highest trees and live in association with epiphytic plants, particularly the orchids and bromeliads. Some ants make their own gardens in tree branches. There, plants sprout from seeds carried there, the roots binding nests and gardens.

Because the life zones of the rain forest are vertical, success goes to those creatures which can move freely up and down. Stranglers thrive in many forests, particularly the strangler figs of Africa, Malaya, and Australia. Seeds germinate high in the forest, usually in the crotch of a branch. Seedlings establish themselves, then send down to the ground aerial roots, which immediately encircle the host trees in strong grips. Then the rapid climb back up into the canopy begins. The strangler, now an independent tree, eventually grows as big as its host.

The creatures of the rain forest show themselves reluctantly, and their secrets are still largely hidden from us. At night, an enormous variety of beetles moves through the tangled undergrowth. Some of them are equipped with twin beams of light shining from their shoulders. They bathe the rain forest with an eerie luminescence. At dawn, the beetles have gone. The countless bats which thrive on forest insects are asleep. The hidden sun casts a pale glow over the gloom to reveal tall trunk spires disappearing into thickets of green. Shiny leaves catch glints of the new light, and young flowers touch the highest limits of the foliage. Below, drooping blossoms hang from limp stems. The dark and brooding interior suggests the mystery of a way of life that has changed little in 60 million years.

141. Stretches of rain forest are patched around the Temperate Zone, but it is the heat of the tropics, high humidity, and a long rainy season that sustain the vast jungles of South America, Africa, and Asia. The edge of the Orinoco River in Venezuela displays the dense tangle of vines and trees typical of a rain forest area.

142. Parrots of the rain forest live in the high canopy and are usually hidden by the dense vegetation. Rivers cutting through the vegetation open up the forest, reveal the sky overhead, and make visible flocks of brightly colored, screeching birds flying to a new food source. Here a golden conure drinks from a bromeliad in the Brazilian rain forest.

144. The fearsome horns of a male Jackson's chameleon seem to have no use except intimidation. This prehistoric-looking monster, resting on a palm branch in the Congo, is a foot long. Its slow movements allow it to stalk insects until it can dart out its long sticky tongue and snare them.

148. The tongue of a brown tree snake in New Guinea flicks back and forth to guide the reptile through the dense vegetation in its nocturnal hunt for small, sleeping birds. The massed branches and vines and the great height and density of the rain forest make it an ideal hunting territory for venomous snakes.
150. Chimpanzees are normally the noisemakers of the Congo, screeching, hooting, and drumming on the trunks of trees to drive intruders away or to announce the discovery of food to the troupe. Here, a young chimp has just discovered the body of its dead mother.

146, 147. A skipper butterfly, a shield bug, a long-horned beetle, and a Coreidae beetle form a vignette of the diversity of insects that have evolved around the rich flora of the tropical forest. Here, where the seasons never change and the food supply is constant, more varieties of insects can be found than anywhere else on earth.

10 The Frontier Mountains

Photographs by:
Ed Cooper, 155
Willis Peterson, 156
Galen Rowell, 158
René Pierre Bille, 159
Leonard Lee Rue III, 160
George Silk: Time/Life Picture Agency, 161
Maurice Hornocker/Photo-West, 162, 163
Charles Steinhacker, 164

Many years ago, high in the New Zealand Alps, I framed a chamois in the telescopic sights of my rifle. Poised and confident, the goatlike antelope stood alone, glowing in the saffron light of a setting sun. It looked toward me, suspicious, but was unable to pick me out against the sun. Then the beautiful animal jumped—one agile, vaulting leap that took it more than twenty feet to the edge of what seemed an almost vertical rock face. With another leap, it disappeared. When I walked to the lip and looked down, I saw that the chamois had jumped across a ten-foot wide chasm. I would have needed at least half a day of hard climbing to reach the other side of the chasm. I turned to leave, the air cold as marble against my face and utterly silent.

Space is not a final frontier; the high mountain is also a goal of life. We can climb it. We can put flags on its summit. But we cannot live at its peak, and none of our technology is much use in making it as accessible as the poles, or even the middle of the Sahara. Yet the creatures of the mountains have easily penetrated these high places.

The facts about mountains that stick in my mind have nothing to do with great scenery or heart-stopping falls from precipices. Instead, they are concerned with the surprise of jumping spiders living at high altitudes in the Himalaya, with the gaggle of geese which have been seen flying over Mount Everest at a height of some thirty thousand feet, and with the force of a gust of wind on Mount Washington, in New Hampshire, which was registered at a speed of 230 miles an hour. The mountains are not only barriers to men; they are also monumental obstacles to the winds of our planet. When winds strike the slopes, their force is compressed and flung upward at greater and greater speeds until hurricanes blow at the highest peaks.

The North American black swift, endlessly circling its mountain crags in summer, may be flying far higher than any winged insect could reach, but plentiful insect food is brought to the bird by lowland winds made powerful during their ascent of the mountains. The strong winds also bring pollen and the comatose or dead bodies of insects to other creatures waiting on the tallest peaks for the arrival of the food. Glacier fleas, which cannot fly, survive on wind-blown pollen and the bodies of the insects. The jumping spiders eat mountain springtails—tiny wingless creatures that flick their tails to hop—and small flies, which, in turn, eat rotting vegetation and fungi.

These surprising examples of survival are part of the mosaic of life on high mountain slopes and peaks. The creatures share their isolated world with the lichens, which climb ahead of all other plants, spreading their rich reds and oranges and yellows over bare rocks. The prime function of the lichens is to break the rocks down into soil, so that in the infinity of time there will be opportunities for other plants to take root. Some lichens, though, are eaten by a high-ranging hunter, the Rocky Mountain goat, itself insulated against the hostility of the wind and the cold by its double coat of fur. As the goat works its way methodically across the surface of a rock, snipping off low growths of lichens, it demonstrates the great ability of creatures to live at the limits of possibility. The goat has four stomachs, enabling it to exploit every scrap of nourishment from tiny sprouts of grass, morsels of dried weeds, twigs, and the foliage of stunted shrubs. It changes coats with the season, shedding its shaggy winter wool in the spring for the summer protection of thin hair. Part of its survival technique is in conserving its energy. It suns itself in sheltered places for long hours where it can watch for the approach of any climbing hunter.

In even more remote places, where crags and precipices shape the mountain, the mountain sheep is a model for all other animals living at high altitudes. Its sure-footedness is almost unbelievable; it can work its way up nearly vertical cliff faces because its rubbery, cloven hoofs are concave and sharp along their edges. They allow the sheep to get a grip on a ledge narrower than its feet, and act as suction cups when the sheep is on a sloping surface.

Each animal in the high solitudes possesses physical characteristics well fitted to out-compete most animals seeking entry to their world. The guanacos, which live at extremely high altitudes in the Andes, purify their blood in the thin air with less oxygen than valley dwellers need. Mountain meadow voles dig deep under the heavy snow to find shelter from the blasting wind and cold, and are able to outbreed their periodic population disasters. The mountain pika, a rodent three times the size of a meadow vole, cuts summer grasses and weeds, then spreads them in sheltered places to dry into hay. It winters in an underground nest stocked with this food.

Mountain lions, snow leopards, Siberian tigers, and many small cats all climb mountains in dangerous pursuit of the defenseless creatures living on the heights, yet the hunters cannot stay up there for long in the numbing cold. And they are not nearly as hardy as the plants that live at high altitudes.

An exceptional unity exists between these plants and creatures. The cushion pink, a mountain plant, builds up such a thick tangle of stems that it traps and holds the sun's heat. Insects retreat into this plant solarium on freezing nights or during hurricane winds.

The limber pine is a conventional tree in the foothills, but in the mountain tundra which begins beyond the tree line, it becomes a dwarf on which animals browse. The snow willow grows luxuriantly in the Rockies, yet never rises more than about two inches above the ground, providing food for grazing animals. Most mountain plants wait to bloom until they have conserved a store of energy. Few of them flower before they are three or four years of age. Some wait more than twenty years. They use devices to trap heat, repel wind, and reduce evaporation. Their leaves are dark green and thick; sometimes they are furred, or impregnated with wax.

A mountain caribou, gifted with big, flat feet that tamp down the deep snow, can wander the slopes to browse and root for food. When such a huge creature clips a single flower from a plant, the plant usually dies, since all its life force was invested in that one blossom. But the mountain caribou cannot compete with the plants in climbing ability. Its large body would lose too much heat in the great winds, and it could not find adequate shelter.

The pocket gopher, which has reached to the limit of the mountain tundra, energetically tunnels through thin soil to eat plant roots, thereby killing the plants. The gopher throws up mounds of excavated soil, which the ferocious winds sweep away. Other plants—not to the gopher's taste—move in, and the gopher is forced to look elsewhere for food. The new plants build up humus, making it possible in time for the original plants to return. The gopher comes back to eat them, and the process of destruction and reconstruction is repeated.

Soaring into the clouds and plunging into valleys, mountain ranges compress many different environments into relatively small regions, at least when measured horizontally. In tundra heights, ptarmigans and pipits may be living less than a mile away from meadowlarks in a sheltered valley, or from curlews prowling a nearby shoreline. The boreal chickadee, which inhabits firs flanking the flower-studded meadows that lie between tree line and tundra, flies within the sound of the fox sparrow singing lower down the tree-clad slopes.

Like most lowlanders, I have always delighted in the great sweeping vistas of the mountains, but with the knowledge that my human eyes are sadly fallible for such watching. The high mountains demand piercing vision. Condors can see a rabbit at nearly five miles, and divine whether it is dead or not. Bald eagles can easily pick out a scampering meadow vole at two miles and keep its position marked as they plane down towards it. With their hollow bones and uncanny capacity to use updrafts, the vultures and great eagles are perfect residents of the mountain ranges. Many hawks and owls are also experienced mountain hunters, and they are able to find food where I see only a desolate pinnacle of rock.

I remember once climbing Kilimanjaro, an easy enough tramp at the start, as long as your heart is strong and you can stand the thin air. From Amboseli, on the northern side of the mountain, Kilimanjaro stands like a monumental backdrop for the familiar African animals—lions, elephants, and antelopes. But as I ascended, the vegetation and the animals changed; not only was

I moving upward, I was, in effect, traveling south toward Antarctica. For every thousand feet I climbed, the temperature dropped about five degrees. Man has attempted to conquer the mountain; the Chagga people intensively cultivate the slopes between four and six thousand feet, where the mountain catches enough rain to grow crops. But then the forest, and the animals, take over again. The leopard and the elephant, the antelopes and the monkeys, rule the slopes of Kilimanjaro toward its twin summits. At about ten thousand feet, the forest has thinned and tundralike country, called heath land, begins. Ahead of me lay mist-laden alpine country, where giant shrubs sprouted, gnarled and twisted trees appeared to fight their way to the summit, and flowering plants laid down carpets of blossoms. The whistling cries of rock hyraxes sounded warning of my approach and reminded me of the whistles of marmots in the American mountains.

Then, at fourteen thousand feet, ice and snow took over, and the animals dropped behind me. When I looked down and away, I could see vultures circling the top of a green forest. The Equator was only two hundred miles away, but it was freezing cold, and that mountain wind was blowing. Near twenty thousand feet it was difficult to walk more than a hundred steps without stopping for a breath. My heart was pounding heavily. Abruptly, a small animal appeared in the snow, cried out a warning, and disappeared among some rocks. There may indeed have been an unwise leopard frozen into this world, as legend has it, but I could climb no further that day. Above me, I was sure, small creatures survived in the cold and isolation. I kept imagining I could hear the tinkling cries of a bird. I remembered reading that insects had reached near the peak of Mount Everest, some ten thousand feet higher than I was now. While I rested, deeply chilled in that icy, endless wind, I reflected that I had climbed far above the rest of humanity. But in comparison with the creatures of the mountain, I was an insignificant traveler of the most transient kind.

155. A chaotic spread of peaks in the Olympic Range of the far Northwest of the United States caps lowland and foothill rain forests, which offer a special environment for a rich diversity of creatures and plants.

156. Both the Dall's sheep and a thin growth of grass have reached the high rocky peaks of remote Alaskan mountains. There, footholds inches wide give the sheep a precarious track through the mountains. The great curved heavy horns of the old rams are the weapons with which these animals test each other during the mating season.

158. The mountain goat of western America treads steep cliffs on broad, two-toed feet with sharp-edged, pliable pads. This agile, wary animal ascends the high slopes of the northern Rockies to Alaska. It makes big leaps but sometimes falls.
159. Wild mountain goats climb to dizzying heights. This conquest of the high places is exemplified by the ibex, a grayish-brown animal with spectacular ridged horns and a scraggly beard. It is found from the Alps to the Himalaya and is adept at making leaps among loose stones at the edges of precipices.

160. A cautious mountain creature, the hoary marmot of the alpine areas of northwestern North America basks in the sun. Its loud whistle, which echoes back and forth among the responding rocks, is the first signal of danger for many alpine animals. It resembles its lowland cousin, the woodchuck or ground hog, in its stockiness and coarse-haired coat.
161. The golden eagle is usually seen winging back and forth on thermal currents, its telescopic eyes probing for the small animals on which it preys. The golden feathers of its crown and nape catch the bright sun as it hunts the wild mountainous country of the northern hemisphere.

162, 163. The mountain lion, wary and solitary, is rarely seen by man. A prime deer-killer, this cat ranges from Canada to South America. But, persecuted everywhere, it survives today in the United States only in wilderness areas of the West and a few places in Louisiana and Florida. Six to eight feet long, its value as a controller of deer populations is belatedly being understood.
164. The native mountain sheep of North America feeds on grasses, sedges, and heathers in its northern habitats. In winter, seen here in Yellowstone National Park, it roams snow-covered ridges scratching out remnants of dried grass. This sheep survives from the mountains of Alberta to northern Mexico.

11 The Fragile Tundra

Photographs by:
Fred Bruemmer, 169
Steven C. Wilson, 170
Co Rentmeester: Time/Life Picture Agency, 172
S. D. MacDonald, 174, 176, 178

In January, the prolific lemmings are hiding under the featureless expanses of Arctic snow, secure in their tunnels from the cold and the ferocious winds. But above them arctic foxes pad softly, crunching the snow, until they smell the lemmings. Raising their tails like flags, the foxes dig in a fury of jabbing paws, their jaws chopping into the snow until the lemmings are obliterated.

Overhead, ravens circle, watching for lemmings. Ermines raise their delicate muzzles high to sniff the air, then disappear in a swift rush into the snow. Before they are eaten, a lemming chorus squeaks alarm. Snowy owls wait on icy ridges for the lemmings, or an arctic hare. Wolves moan in the distance. A polar bear heaves a glacial boulder aside and scoops up the exposed lemmings as they flee frantically.

This is the winter of the tundra, immense desert lands that circle the Arctic. There, despite blizzard, ice, and horrendous winds, is hosted an extraordinary gift of life for the polar region, awaiting the touch of the spring. Five million square miles of trackless tundra make up about one-tenth of the earth's land surface, and while the tundra gets only about eight inches of rain a year, its huge size enables it to support some of the most spectacular populations of animals anywhere.

The first time I saw tundra, I was heading for the Arctic in a Canadian airplane. I have never forgotten the hours it took to fly over the flat, greenish, water-blotched landscape that appeared to have no boundary. But I was not intimidated, as I had been in hot and hostile deserts, nor did I feel the tedium imparted by rain forests. The tundra appeared both benign and beautiful to me as it marched north to its eventual meeting with fields of Artic ice.

I came to know it well in later years. It is a place of magic loveliness, where black winter air sometimes seems to catch fire, and showers of color fall from high places. These subtle violets and purples of the aurora borealis drop slowly while sparkles of green suggest forests on the other side of the earth.

I once saw the colors of the northern lights catch for a moment in the dense and bushing white fur of a night-prowling fox, so that the animal became luminous in the snow. I had seen the fox trailing a polar bear in the hope that the bear would eventually find a seal hole and leave scraps of meat from its kill for the fox.

Unlike the Antarctic, where animals have secured only a tenuous foothold on the continent, the creatures of the tundra and the Arctic have made the entire territory completely their own.

The prudent fox has caches of frozen lemmings buried in a score of places. It also has refrigerated stores of duck eggs, ducklings, and shore birds—any meat it has been able to hoard during the brief tundra summer. The half-ton musk ox, sedentary and thickly furred, stands exposed for days in blizzards, waiting for the wind to blow snow from ridges so it can graze on the frozen remains of last summer's plants.

The musk ox stands fast to beat the winter, but the polar bear keeps moving. It is the one creature which truly bridges the ice pack of the pole and the frozen or water-logged tundra surrounding it. Bears wander near the pole. They dig seals from ice caves and swat them dead at breathing holes. They cross the ice to land, roam through lemming country, ride icebergs down the coasts of Greenland.

In delicate counterpoint to the powerful, nomadic bear are the many animals which have made their own, unobtrusive conquests of the north. I once watched a pair of glaucous gulls soaring over the tundra like vultures, silent and graceful and almost invisible against the white winter sky. The delicate ivory gull is a beautiful, almost dovelike creature with red eyelids and black feet. It hovers in the wind and does indeed disappear in midair, so well does its plumage match the white sky. And when a snowy owl glides overhead, it is invisible, except for its eyes. I have seen a ptarmigan explode from a snow bank, white on white, and disappear.

The tundra is a tenacious interlocking of lives, each one totally dependent on some other life. The ivory gull scavenges at the bear's feast and follows seals a thousand miles south when they migrate to breed. The caribou are not as tough as musk oxen, and they retreat south in winter to crop lichens buried in the snow. The arctic hare has grown to be twice as big as the fox, and so it is safe from what should be its natural enemy. In response to the bitter cold, the fox has diminished in size to about eight pounds, so that its small, lithe body needs only a minimum amount of food to sustain it.

To this frozen, tough, expansive world comes a spring which has no counterpart anywhere. It is twenty years since I experienced one, but the memory remains sharp, brilliant, unchanged. By April the dawns are rapid, turning quickly from gray to pink to gold, each day the sun rising higher in the sky. By late May the sun is continuously above the horizon. The light seems to magnify a man's vision so that distant objects appear to be close.

On the day before the thaw began, total silence gripped the land, except for snow buntings twittering overhead in early expectation of finding bare ground. Strings of snow geese zigzagged in search of open water. The thaw was announced by the tinkling of a distant bell—the movement of slowly trickling water. The bell was joined by many others until thousands of them rang in a full-throated boom as water rushed through a disintegrating winter world.

Now there were only fourteen days before summer was upon the tundra. The Arctic air was decorated with legions of newly arrived birds. Hovering and singing in flight, displaying and squabbling for territory, the snow buntings, longspurs, pipits, larks, and millions of shore birds claimed their mates. Arctic poppies bloomed immediately. Purple saxifrage blossoms spread. Dwarf willows gleamed with fresh leaves. I was overwhelmed by the flush of restored life.

In this exploding world the caribou were moving. I followed them for two weeks with a group of scientists. The caribous' heedless, splay-footed legs, so ideal for walking on thick snow, now crunched across the quaking surface of new bogs, crushing many of the nearly one thousand species of plants that have colonized the tundra.

They walked among the dwarf forms of trees which, like the arctic fox, were miniature versions of their southern relatives. There, a six-inch-high willow might be four hundred years old. The caribou plodded across thawing masses of sodden sphagnum moss, plowed through meadows of arctic cotton, descended into sheltered hollows where heather and blueberry and cranberry have kept precarious hold in the north.

The march of the caribou was not only a migration to expanding pastures; it was also a flight from the billions of biting insects that were rising from their winter sleep as pupae in the snow. Mosquitoes, blackflies, butterflies, moths, bumblebees—and countless other insects—poured out into the clear, cool air. Only insects can sleep through the winter in the tundra. During the brief summer, larger animals have no time to put on the fat they would need to hibernate through the long and severe tundra winter.

The caribou were overtaken in their march by legions of these new arrivals. They tramped along the shores of tundra lakes and ponds where old squaw ducks were nesting, along with baldpates, teal, pintails, scaups, scoters, mergansers, and goldeneyes. The voices of plovers and sandhill cranes filled the air. Gyrfalcons sped to fatal collisions with frantic ducks and fleeing ptarmigans. Circling ravens uttered occasional melancholic croaks. Long-tailed jaegers paused, their gracefully elongated and dusky bodies capped by black-topped heads and white chests. Their webbed feet were reminders that these creatures, now hundreds of miles from the sea, had made a special adjustment to this tundra world. They dropped to seize lemmings, and pounded them to death against the wet ground.

This was not a big lemming migration year, but enough of the little creatures were abroad to interest the caribou, and where the lemmings had gathered to catch the sun, making a thick carpet of their prone bodies, the caribou nuzzled forward to eat them as if they were plants. Had this been a big year for lemmings—one of their periodic population explosions—then owls and falcons, gulls and jaegers, foxes and wolves, bears and ermine would hunt them down in shrieking, squabbling congregations.

In such fecund years, the bounty of

169. The tundra, vast and treeless, has a frozen subsoil and is covered only with such low-growing vegetation as mosses, lichens, and stunted shrubs. It forms a zone around the earth from the Arctic ice cap south to where trees begin to grow.

lemmings is passed on immediately. Arctic foxes, fat on lemming flesh, double the size of their litters. Equally replete snowy owls lay nine eggs instead of four, while gulls, hawks, falcons, and ravens also increase their numbers. Tundra creatures seize the briefest of chances, and multiply whenever they can.

The tundra is so vast, however, that its creatures are only occasionally gathered into large groups. Caribou, which were once seen in the millions, now migrate in thousands. I once flew over an odd congregation of about one hundred arctic hares, incongruously white against the saxifrage meadow, and saw no sign of any other hares for hundreds of square miles around the isolated group. Musk oxen may gather in groups, and form circles to protect themselves against wolves, but some can wander alone for years, presumably never meeting wolves.

Across the vastness of the tundra the animals are spread thin, and they often appear amazingly tame to human beings. Two cock ptarmigans once fought at my feet for the possession of a scrap of territory in an empty landscape, and they did not seem to be the least bit alarmed by my presence. An old squaw duck waited until I was only ten feet away from her before she flew off. A sanderling, one of the millions of shore birds which use the Arctic tundra to nest, sat on the eggs of his mate so tightly that when I picked him off he never moved. I set him down again.

An arctic wolf, all legs and subtle grace, watched me with somber and unafraid eyes as he fed off the remains of a musk ox carcass.

The Arctic summer dies a quiet death. On late August afternoons the air is touched with a gentle hint of winter; it feels like a cool hand near your face. The pale sun, gold as an old doubloon, sits on the horizon while nearby rocks are held still in mauve mist. This is the time when the tundra's fruits ripen.

In places there is such an abundance of cloudberries, crowberries, bilberries, and cranberries that the juices of these fruits pass through the digestive systems of birds and mammals to splash the tundra with their bright colors.

The tinkling shore birds disappear. The caribou have turned back toward the southern timberline. Young snow geese test their juvenile wings. King eiders molt, and clumsily splash across ponds on new flight feathers. The standfasts —the hares, foxes, ermines, ptarmigans, musk oxen, wolves, bears, and lemmings—brace themselves for the cold that lies ahead.

But sometimes during the short summer the sun thaws out part of the tundra permafrost deeper than usual over hundreds of square miles, and then large sections of the tundra moves downhill wherever there is the slightest slope, shearing off the roots of all the tenuously established plants. Where plant growth has been too heavy, however, the sun does not warm the ground at all, and the permafrost thickens. The plant roots are doomed to freeze prematurely and die. The tundra's shallow crust of organic matter is so thin that even tiny changes in its condition mean that life may not thrive there for centuries. The fragility of the tundra can teach the perceptive watcher that neither creature nor man can survive for long if the earth's thin layer of soil, upon which life depends, is broken or destroyed.

170. In the fall, three caribou, part of a vast herd of migrating animals, travel south from Alaska's North Slope along a regular route to escape the harsh Arctic winter. They may cover as much as 500 miles in a season.
172. In early winter, a polar bear mother and her three grown cubs will soon separate. Male polar bears spend the winter wandering the ice pack hunting seals. The females, especially when pregnant, seek out snowy areas on land and go into semihibernation. The cubs will be born in late winter or early spring.
174. The arctic wolf differs from the timber wolf only in its whiter and thicker pelage. Arctic wolves not hunted by man are unafraid and curious and will come within a dozen feet of a human intruder. This wolf, seen on Bathurst Isand, hunts Perry's caribou.
176. Bull musk oxen on Bathurst Island, interrupted in their feeding, begin to move together to take up their circular defensive position. Once they are back-to-back, they cannot be attacked from the rear.
178. This pure white ivory gull, adapted to life in the high Arctic, will travel long distances in the track of an arctic wolf or a polar bear in order to feed on the leavings.

NOTES ON THE PHOTOGRAPHERS. The numbers in parentheses refer to pages.

Anthony Bannister (118, 122) was taken to South Africa as a child and soon became interested in invertebrates and photography. After publishing over a thousand pictures and many articles, he gave up a career in electronics to become a photographer of natural history.

Jen and Des Bartlett (90) are an Australian camera team famous for the wildlife films and stills they have made in the past twenty years on every continent. Among their best-known films are *World of the Beaver* and *Flight of the Snow Goose*.

Erwin Bauer (128) is an adventure and wildlife photographer based in Jackson Hole, Wyoming. He is author-cameraman of *Treasury of Big Game Animals* and *Hunting with a Camera*.

René Pierre Bille (159) was born in Switzerland and lives in a high mountain village. He is known for his photographs of wildlife and his books on this subject, the latest of which is *Mountain Animals*.

Les Blacklock's (82) early years in the Moose Lake, Minnesota, area turned him to a career as a wildlife photographer. He has had photographs in many national magazines and has collaborated with author Sigurd F. Olson on *The Hidden Forest*.

Dennis Brokaw (34) graduated from the University of San Diego and became a flight test engineer, but he later turned to photography. His work has appeared in many leading publications.

Bill Browning (84), free-lance photographer/writer/artist, has numerous credits in national magazines and outdoor books, has produced television movies, and is currently producing wildlife fine art.

Fred Bruemmer (169) worked in Canada as a newspaper photographer and reporter before becoming a free-lancer specializing in the Arctic. He has written 300 articles and four books, including *The Arctic*.

Robert Carr (131) spent his early years "chasing animals and climbing trees" in Michigan. He turned to photography in order to continue chasing animals. He has appeared in *National Wildlife* and other magazines and books and teaches environmental photography at Lansing Community College.

Patricia Caulfield (72) graduated from the University of Rochester. She joined the staff of *Modern Photography* and became executive editor but resigned in 1967 to devote herself to nature photography and conservation. She created an outstanding Sierra Club book on the Everglades.

David Cavagnaro (42, 46) studied entomology. After winning a *Life* photo contest in 1970, he carried out photographic assignments for Time-Life books. He is the author of *This Living Earth* and is resident biologist at the Audubon Canyon Ranch in California.

Glenn D. Chambers (88) is a biologist and wildlife photographer for the Missouri Department of Conservation. His photographs have appeared in Time-Life books and his films on wildlife subjects have been widely distributed.

Neville Coleman (22) is an Australian underwater naturalist who recently completed a survey of Australian marine life. His articles and photographs appear in many publications and he is author of *Australian Marine Fishes in Colour*.

Ed Cooper (117, 127, 155) was brought up in New York but then moved to the West Coast and took up photography. His pictures appear frequently in many wilderness and conservation publications.

William R. Curtsinger (28) is a contract photographer for the *National Geographic* specializing in natural history and underwater photography. He is also photographic coordinator of Project Jonah, an organization to save the whale.

Thase Daniel (73, 74) was born in Arkansas and after studying music turned to photographing the life of the bayous. She has since photographed animal life from the Amazon to the Arctic and has contributed to many wildlife magazines and books.

Edward R. Degginger (37, 96, 108, 144) is both a professional chemist and a photographer of wildlife. Over 2,800 of his pictures have appeared in exhibitions, books, magazines, and encyclopedias.

Jack Dermid (70) is an editor of *Wildlife in North Carolina* and associate professor of biology at the University of North Carolina. His wildlife photography has appeared in many magazines and books.

Tui De Roy's (1, 40) family moved from Belgium to the Galapagos Islands when she was a baby. Her unique opportunity to explore the islands' animal and plant life has resulted in articles and photographs by her in *Audubon* and other publications.

John Dominis (104), one of the most versatile photographers of our day, studied cinematography at the University of California and played football in the Rose Bowl. After working for various magazines he joined the *Life* staff in 1950. As a wildlife photographer he is best known for his series in *Life* on the "big cats" of Africa.

Hans Dossenbach (109) was born in Switzerland and studied animal behavior in Vienna and photography in Zurich. He is the author and photographer of twelve books, including *The Family Life of Birds*.

Douglas Faulkner (19, 26, 142), internationally known underwater photographer, studied marine biology and then began his career with a trip to the South Pacific in 1962. His photographs have been featured in leading publications here and abroad. His most recent book is *This Living Reef*.

Jean-Louis Frund (134) is a French-Canadian whose main interest is nature. He photographs for various Canadian governmental agencies and for the Musée National des Sciences Naturelles.

Jessie O'Connell Gibbs (38), a native of Charleston, South Carolina, began her career as a newspaper photographer. Her photos, especially of shore birds, have appeared in *Life* and other publications.

Fritz Goro (113, 48) studied at the Bauhaus and worked in Germany as an art director. He emigrated to the United States in 1936, soon started working for *Life*, specializing in science photography, and remained with it until it closed. His work has appeared in hundreds of books.

Clem Haagner (110, 121) is a South African poultry breeder whose hobby is photographing wildlife. His work has appeared in many publications around the world and he is now preparing an illustrated book on the Kalahari Desert.

Bob Harrington (98, 101, 102, 112) has since 1950 been a photographer in the Michigan Department of Natural Resources. He spent eighteen months in Africa teaching wildlife management. His photos have appeared in *Life, National Geographic,* and other magazines.

George Holton (41, 43, 56, 58, 60, 100, 106, 150) is a New York photographer specializing in natural history and archaeology. Photographs taken on his travels to remote places, including eight trips to Antarctica, have appeared in many publications. His latest book, on New Guinea, is *The Human Aviary*.

Maurice Hornocker (162) began photographing wildlife and wilderness while at the University of Montana. His photographs, especially of western mammals, have appeared in national publications.

Eric Hosking's (135) photographs of nature and wildlife have apeared in over 800 books since 1929. His autobiography, *An Eye for a Bird*, was published in 1970. He also co-authored *Wildlife Photography*.

Philip Hyde (33) studied photography under Ansel Adams and Edward Weston. He has illustrated seven books, including *Slickrock* and *An Island Called California*, and has appeared in many publications.

M. Philip Kahl (76) earned a Ph.D. in zoology at the University of Georgia and has studied the storks and flamingos of the world mostly under research grants from the National Geographic Society. His photographs have appeared in *National Geographic* and other publications.

Hubertus Kanus (95) is a German photographer and journalist with a special interest in Third World countries. His work has appeared in countless magazines and books here and abroad.

S. D. MacDonald (44, 136, 174, 176, 178) is curator of vertebrate ethology at the Museum of Natural Sciences in Ottawa. His pictures have appeared in *Nature Canada, Life, Audubon,* and elsewhere.

Lorus and Margery Milne (36) are professors of zoology at the University of New Hampshire. Their photographs of animal and plant life have appeared in national magazines and books. They are also the authors of more than thirty books on these subjects.

Willis Peterson (156) has been photographing na ture since he grew up in Colorado Springs. He ha been host of a television series featuring his natur films, and 130 of his photographs are now on tou He directs the Photographic Department at Glendal College in Arizona.

David Plowden (81) studied economics at Yale Un versity and then photography with Minor White i Rochester. He has published nine books and appeare in many magazines.

Hans Reinhard (130) heads the photo laboratory i the Orthopedic Clinic at Heidelberg and he ha spent much time photographing animals from th nearby Odenwald. He is the author of *Die Techni der Wildphotografie*.

Co Rentmeester (172) came to the United Stat from Amsterdam in 1941. He joined *Life* in 196 and did not only leading news stories but also evoca tive nature photographs. He was named Magazin Photographer of the Year in 1972.

Edward S. Ross (147), curator of entomology at th California Academy of Sciences, pioneered in cand insect photography in his *Insects Close Up* (1953 He has since done every type of nature photograph in almost all parts of the world.

Galen Rowell (158) has made a hundred ascents the mountains of western North America ar Alaska. His articles and photos on wilderness su jects have appeared in *National Geographic*, Sier Club magazines, and elsewhere, and he is author *The Vertical World of Yosemite*.

Leonard Lee Rue III (160) has devoted his life wildlife writing, lecturing, and photography. H articles and photos have appeared in 400 publicatio and he has published many books on the animals North America.

Emil Schulthess (53), famous Swiss photograph and author, has won many photographic award His pictures have appeared in leading magazines over the world and he has done books on the Amaz and Antarctica. He is now making helicopter pan ramas for a book on the United States.

George Silk (44, 161), born in New Zealand, serv as a war correspondent-photographer in 1939–4 He was a *Life* staff photographer from 1943 to 197 His nature and wildlife photographs have appear in Time-Life books and elsewhere.

Gordon S. Smith (65) is an Englishman now livi on Cape Cod. His work appears in many magazin and books.

Charles Steinhacker (164) directs the Rangel Lakes Photography Center in Maine. He has pu lished *Superior: Portrait of a Living Lake* and w the main photographic contributor to *Yellowstone A Century of the Wilderness Idea*.

Harald Sund (6) lives in Washington State a spends much of his time exploring and photograp ing the surrounding mountains. His work has a peared in several Time-Life wilderness books.

Karl H. Switak (120, 148) is supervising her tologist at the California Academy of Sciences. has published articles, with photographs, in *Nation Wildlife*, and a book, *Desert Reptiles*.

Valerie Taylor (16, 18, 24) is an Australian div who, along with her husband Ron, specializes underwater photography. Their television seri *Inner Space*, was widely distributed and she w featured in *Blue Water White Death*. Her stills ha appeared in *National Geographic* and elsewhere.

Karl Weidmann (141, 146) grew up in Switzerla and then fulfilled a boyhood dream of roaming tropical jungles by moving to Venezuela. There photographs wildlife and Indians. His work appea in many media and his book *Faraway Venezuela* w be published soon.

Larry West (66, 68, 132) is a nature photograph whose pictures have appeared in many books a magazines. He lives in Mason, Michigan.

Steven C. Wilson (170) majored in fisheries at t University of Washington and now lives in Was ington. He attempts through films (he has ma seven) and widely published photographs to " crease man's awareness of the other living thin on our planet."